4th International Conference on Advanced Sciences 2017

IOP Conference Series: Materials Science
and Engineering Volume 464

Hurghada, Egypt
7 – 10 November 2017

ISBN: 978-1-5108-8009-2
ISSN: 1757-8981

Printed from e-media with permission by:

Curran Associates, Inc.
57 Morehouse Lane
Red Hook, NY 12571

Some format issues inherent in the e-media version may also appear in this print version.

Printed by Curran Associates, Inc. (2018)

For permission requests, please contact the Institute of Physics at the address below.

Institute of Physics
Dirac House, Temple Back
Bristol BS1 6BE UK

Phone: 44 1 17 929 7481
Fax: 44 1 17 920 0979

techtracking@iop.org

Additional copies of this publication are available from:

Curran Associates, Inc.
57 Morehouse Lane
Red Hook, NY 12571 USA
Phone: 845-758-0400
Fax: 845-758-2633
Email: curran@proceedings.com
Web: www.proceedings.com

TABLE OF CONTENTS

IOP Conf. Series: Materials Science and Engineering **464** (2018) 011001 doi:10.1088/1757-899X/464/1/011001

Editorial note

Prof. A.A. Farghali

Assoc. Prof. S.I. El-Dek

In this year, we celebrated the ranking of our university "Beni-Suef University" as the 1st Egyptian University, 601-800 world wide and 301-400 in the physical sciences, 101-150 in the young universities as Times Higher Education published. It is also a great pleasure to announce that for the 1st time we are included at Shanghai ranking (901-1000). We are willing to compete next year on advanced positions.

Faculty of Postgraduate studies for advanced sciences (PSAS) is an exclusive faculty, a 1st class learning place that relocates you from multidisciplinary, interdisciplinary to transdisciplinary research in FOUR main departments working collectively to achieve the best prototypical education on both regional and international level.

PSAS is distinguished by an expertly qualified team working together with a large number of research facilities in the specified laboratories.

It is gratifying to note that the agenda of the conference covers a wide range of very interesting items related to the nanotechnology, biotechnology, environmental sciences and all energy related subjects and especially those directly related to aspects of all advanced sciences.

We are aware of the tremendous effort made, by all faculty members from day 1 up until now.

No matter how much we can do by ourselves on the national level, whether it be research or development, it is never enough. In a spirit of true cooperation, we were proud to announce for the first time reviewing and publishing international proceedings in IOP conference series: materials science and engineering. The editorial board of this issue Dr.A.A.Farghali and Dr.A.H.Zaki and Dr.S.I.El-Dek.

We can, and must, track PSAS as a multi-dimensional concept, encompassing the economic, social, institutional and physical elements of the excellent research educational platform, in a wider sense, it would be relatively easy to affect the necessary adjustments for a truly effective cooperation on the regional and international basis. This is fully consonant with our official position taken and the full support of the concept of internationalization

IOP Conf. Series: Materials Science and Engineering **464** (2018) 011001 doi:10.1088/1757-899X/464/1/011001

This special issue on "Materials sciences and engineering" resulted from the 2017, 4th International conference on advanced sciences ICAS4, held in Hurghada, Egypt on 7-10 th November, 2017. This international conference is the 4th one organized by our group at the Faculty of postgraduate studies for advanced sciences brought together academics, scientists, engineers and postgraduate students to exchange and share their knowledge and research results on all aspects of multidisciplinary sciences and to discuss practical challenges and solutions. The topics covered include:

- Materials Sciences and Nanotechnology
- Biotechnology and life sciences
- Renewable energy science and engineering
- Environmental science and industrial development.

The event was coorganized with Vigo University-Spain and the ENNN (Egyptian national nanotechnology network) and sponsored by many Egyptian industrial communities. Following the submission of abstracts to the international conference, 10 extended full papers have been reviewed from 25 categorized in Materials science and nanotechnology session and chosen for inclusion in this special issue. Collectively, these papers have addressed several key issues. I am therefore delighted to present these papers for all readers.

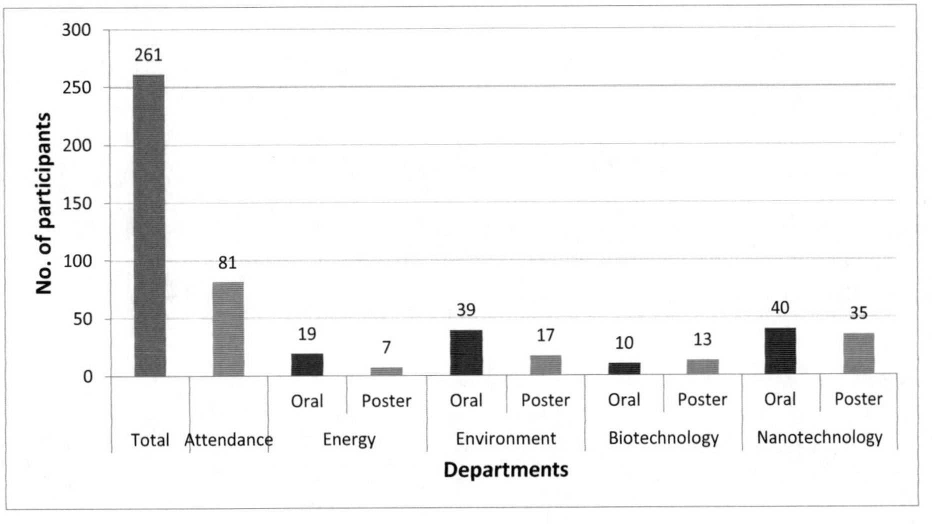
300
250
200
150
100
50
0
No. of participants
261
81
19
7
39
17
10
13
40
35
Oral
Poster
Oral
Poster
Oral
Poster
Oral
Poster
Total
Attendance
Energy
Environment
Biotechnology
Nanotechnology
Departments

IOP Conf. Series: Materials Science and Engineering **464** (2018) 011002 doi:10.1088/1757-899X/464/1/011002

Scientific Committee:

International

Prof. Rasim Mogulkoc (Selcuk University, Turkey)
Prof. Mamoun Mohamed (KTH, Sweden)
Prof. Rashid Ahmed (UTM, Malaysia)
Prof Rafaqat Hussain (COMSATS, Pakistan)
Prof. Nguyen TK Thanh (LCN, UK)
Prof. Kazwichi Hayakawa (Japan)
Prof. Arthur Bros (Netherland)
Prof. Shaker Moussa (USA)

National

Prof. Nadia Zakhary (Cairo university)
Prof. Mahmoud Hashem (GUC)
Prof. Essam Abdel-Hady (Minia University)
Prof. Nageh Allam (AUC)
Prof. Taha Matar (CMRDI)
Prof. Ibrahim M. El-Sherbiny (Zewail City)
Prof. Nagwa El-Badri (Zewail City)
Prof. Dr. Waheed Emam(Ein Shams University)
Prof Dr. Anwar B. Mansour (Cairo University)

Peer review statement

All papers published in this volume of *IOP Conference Series: Materials Science and Engineering* have been peer reviewed through processes administered by the proceedings Editors. Reviews were conducted by expert referees to the professional and scientific standards expected of a proceedings journal published by IOP Publishing.

Detection, ecological risk assessment and removal efficiency of diclofenac and caffeine in wastewater treatment plant

Heba A Younes[1], Hamada M Mahmoud[1,2,*], Maha M Abdelrahman[3] and Hossam F Nassar[1]

[1] Department of Environmental Sciences and Industrial Development, Faculty of Postgraduate studies for Advanced Science, Beni-Suef University, Egypt.
[2] Zoology Department, Faculty of Science, Beni-Suef University, Egypt.
[3] Pharmaceutical Analytical Chemistry Department, Faculty of Pharmacy, Beni-Suef University, Alshahed Shehata Ahmed Hegazy St., Beni-Suef 62514, Egypt.
[*] Corresponding author, e-mail: mhmada@aucegypt.edu

Abstract. Pharmaceutical residues that are considered as emerging contaminants have received increased interest in the last years due to their adverse effects on the environment. This is the first study on the assessment of diclofenac and caffeine in municipal waste water of Egypt. The study aims also to report the removal efficiency of waste water treatment plant and the ecological risk assessment of diclofenac and caffeine. Diclofenac and caffeine were detected in both influent and effluent of Tezmant waste water treatment plant (WWWP) in Beni-Suef governorate, Egypt with detection frequency of 100% by using UPLC-MS/MS method. The maximum detected concentrations were 27.85 μg/L for caffeine in February and 1.07μg/L for diclofenac in January, respectively. Caffeine showed higher removal efficiency (95.2%) than that of diclofenac (69.6%). Based on Predicted No Effect Concentration (PNEC) obtained from the literature, it is observed that diclofenac has high risk (RQ>1) on fish and low risk on algae and Daphnia. Meanwhile, caffeine showed low risk on all of the studied species.

Keywords: Diclofenac; Caffeine; Wastewater Treatment Plants; Removal efficiency; Solid phase extraction; Ecological risk assessment

1. Introduction

Pharmaceuticals are biological active compounds that are very important in our life because of their effects on living organisms even at low concentration for treatment of illness and the reduction of pain. However, they may impose harmful effects on different organs through their side effects. Discharging of Pharmaceutical residues to the environment may induce acute and chronic effects on living organisms [1–5]. Recently, there has been a great concern about pharmaceutical residues in the environment due to their persistency and continuous input [6,7]. They find their way to the environment through many routs such as effluents of municipal wastewater treatment plants (MWWTPs) and landfill effluents.

Among the largest consumed pharmaceutical drugs in Egypt and many countries around the world are diclofenac and caffeine. Diclofenac is non-steroidal anti-inflammatory drug that is used for treatment of several diseases including arthritis and rheumatoid diseases. Caffeine is a natural alkaloid that is extracted from some plants. Caffeine is used increasingly as a stimulant to the central nervous system and it is found in coffee, tea and many pharmaceutical products [8]. Caffeine and diclofenac are detected in waste water and other water masses around the world as well as drinking water [9–11]. Although the detected concentrations are very low (from ng/L to μg/L), their continuous input to the environment induced undesirable ecological effects such as decreases of vulture population in Pakistan [12,13] and human health effects. The presence of diclofenac in drinking water even at low concentration is a very serious problem to humans as it causes renal lesions as well as cytotoxicity to liver and kidney [14–17].

Although there are several studies worldwide evaluating the residual concentrations of diclofenac and caffeine, the scarcity of available information on their quantities in different compartments of the aquatic ecosystems in Egypt as well as their reported drastic ecological and human health effects necessitated the present study. This study aims to detect the residual concentration, removal efficiency and ecological risk assessment of both caffeine and diclofenac in a municipal wastewater treatment plant that serves a large population in Beni-Suef governorate, Egypt. The massive use of Egyptian community of the two investigated drugs also added to the value of the present study.

2. Materials and method

2.1 Chemicals and materials

Diclofenac sodium standard (DCF) (98.74%) and caffeine (CAF) (99.45%) obtained from sigma Aldrich. All reagents used were of HPLC grade obtained from Fisher scientific company (methanol, Acetonitrile and water). Ethylenediamine tetraacetic acid disodium salt dihydrates (EDTA) (99.0–101.0%) and sodium phosphate monobasic dehydrate are obtained from Sigma-Aldrich. Formic acid (analytical reagent grade, 98%), Hydrochloric acid, 36.5-38% (ACS BASIC Scharlau), Phosphoric acid (Laboratory reagent group) were supplied by Fischer Scientific. Oasis HLB cartridge (200 mg, 6 cm3) Waters (Milford, MA, USA) is used. Glass microfiber filter (1 μm GF/B, 0.45 μm) Whatman is used to remove suspended solids and syringe filter (PVDF) 0.45 μm pore size used in filtration of sample before injection to UPLC.

2.2 Sampling

Twenty-four 24-h composite samples were collected from the influent and effluent of Tezmant municipal waste water treatment plant in Beni-Suef governorate, Egypt. Sampling was carried out in winter season (from December 2015 to February 2016). Samples were collected in dark amber glass bottles, then transported to the lab in ice bag and extracted in the same day of collection to reduce the degradation of the target compounds.

2.3 Extraction

The sample was filtered using 1 μm then 0.45 μm glass microfiber filters and acidified with 0.1 M HCL to pH=2, then 0.5 g EDTA was added. Solid phase extraction cartridge was conditioned three times by 5 mL methanol and then three times by 5 mL deionized water for activation. The sample was then passed through the cartridge. After that the cartridge was washed by deionized water and dried for about 20 minutes under a vacuum. Elution process was made by 5 mL methanol twice. The extract was evaporated by using rotary evaporator at 50 ^{o}C then re-dissolved in 2 mL methanol and 1 mL 0.1% formic acid solution (0.1% formic in methanol). The final extract was cleaned by filtration with 0.45 μm PVDF syringe filter before UPLC-MS-MS analysis.

2.4 Analysis

Diclofenac and caffeine were analyzed by using Ultra-performance Liquid Chromatography tandem mass spectrometry System UPLC-MS/MS (waters) with positive ion mode electrospray. The used detector is Triplet Quadrapole. First, the separation done using C18 Column, 3.5 μm particle size (4.6 ×50 mm) which was maintained at 40oC using the mobile phase consisting of water containing 0.3% formic acid and 0.1 ammonium formate (solvent A) and acetonitrile- methanol (in ratio 1:1, v/v) (solvent B) with Positive Ion Electrospray mode at a flow rate of 0.15- 0.30 ml/min at a gradient elution. The elution program started with 5% eluent B for 4 min then increased to 88%, 100% at time 22.5 and 23 min consequently. The injection volume was 15µL. Mass spectrometry conditions are source temperature at 140oC and desolvation temperature 350oC. Quantification was done based on peak areas of target compounds. Standard stock solutions of diclofenac and caffeine (1 mg/mL) were prepared in methanol, and then proper dilution was performed to prepare working solution of both drugs in concentration of (1 µg/mL) in methanol. The calibration curve was done in the range of 0.001-1 µg/L (9 points) for caffeine and 0.01-1 µg/L (10 points) for diclofenac by dissolving them in methanol. The correlation coefficients (R2) of the calibration curves are shown in table 2. The limit of detection (LOD) was determined as the lowest concentration gives a signal to noise (S/N) ratio of 3 and the Limit of Quantification (LOQ) was measured with S/N ratio of 10. The recoveries of the target compounds were carried out by spiking three water samples with standard drugs at concentration of 100, 200 and 300 ng/L (Table 1).

Table 1. Main analytical parameters of the target pharmaceuticals.

Compound	LOD[a] (ng/L)	LOQ(ng/L)[b]	R^2	Recovery %[c]
Diclofenac	2.85	8.55	0.982	58.79
Caffeine	0.31	0.95	0.985	80.97

[a] LOD = Limits of detection,
[b] LOQ = limit of quantitation
[c] Data are expressed as average of 3 determinations.

2.5. Removal efficiency estimation

The efficiency of wastewater treatment plant for the removal of caffeine and diclofenac was calculated by the concentration of them in the aqueous phase in the influent and effluent [18,19] as presented in equation (1).

$$\%\ \text{Removal efficiency} = \frac{C_{influent} - C_{effluent}}{C_{influent}} \times 100 \qquad (1)$$

2.6. Ecotoxicological risk assessment

The expected risk resulted from the discharge of the effluent waste water containing caffeine and diclofenac to the environment was calculated by using Risk Quotient (RQ) which is the ratio of the measured environmental concentration (MEC) to its Predicted No Effect Concentration (PNEC).

$$RQ = \frac{MEC}{PNEC} \qquad (2)$$

Where MEC corresponds to the highest detected effluent concentration
PNEC is the safe dose of compound concentration and it is measured using equation (3).

$$PNEC = \frac{(EC_{50}\ \text{or}\ LC_{50})}{AF} \qquad (3)$$

Where (EC_{50}) is the median effective concentration, AF (Assessment factor) is the safety factor or uncertainty factor which set 1000 as recommended by European Commission [20]. According to Hernando et al. 2006, RQ values will determine the resulted risk as following: RQ is ≥ 1 indicates high ecological risk, $0.1 \leq RQ < 1$ a median risk is suspected and $0.01 \leq RQ < 0.1$ low risk may appear.

3. Results and discussion

3.1. Occurrence of diclofenac and caffeine

Twenty-four samples were collected from influent and effluent of wastewater treatment plant during winter season (2016) in December, January and February. As shown in Figure 1, the lowest detected concentration was in December for both diclofenac and caffeine. Caffeine concentrations were higher than diclofenac in all months.

Diclofenac was detected in all of the samples with maximum concentration of 1.07 µg/L and minimum concentration below limit of quantification (Table 2). The detected concentrations were lower than that reported by Papageorgiou et al., 2016 (4.869 µg/L) and Kosma et al. 2014 (5.16 µg/L) in Greece but also Lower concentrations were detected in Thailand (0.382 µg/L) [21].

Caffeine detection frequency was 100% for all influent and effluent samples and It was detected with relatively high maximum concentration (27.85 µg/L) (Table 2) that was similar to the majority of previous studies on occurrence of caffeine for example , 192 µg/L in Mediterranean Coast [22], 96.65 µg/L in Greece [10], 89.5 µg/L in China [23]. Otherwise, there are relatively low concentrations of caffeine such as that reported by the following research articles: 5.559 µg/L in Greece [3] and 4.550 µg/L in Thailand [21].

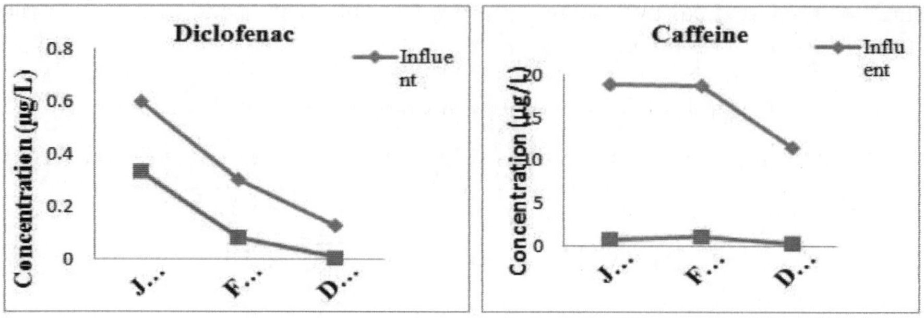

Figure 1. Occurrence of diclofenac and caffeine in Tezmant wastewater treatment plant, Egypt.

The high concentration of caffeine may be due to the high rate of drinking coffee and tea in the Egyptian community.

Table 2. Concentration of diclofenac and caffeine in both influent and effluent of municipal waste water treatment plant.

compound	Influent			Effluent		
	Min (µg/L)	Max (µg/L)	Frequency	Min (µg/L)	Max (µg/L)	Frequency
Caffeine	6.67	27.85	100%	<LOQ	2.32	100%
Diclofenac	0.03	1.07	100%	<LOQ	0.65	100%

3.2. Removal efficiency

There are many parameters that affect the removal efficiency of the waste water treatment plant for the pollutant such as its physicochemical properties that affect whether a compound will remain in the aqueous phase or adsorbed to sludge and wastewater treatment plant operating conditions. Adsorption onto suspended solids, aerobic, and anaerobic biodegradation, and chemical degradation (via processes such as hydrolysis) and volatilization are the primary removal mechanisms for pharmaceutical residues in waste water. As shown in Figure 2, Diclofenac removal rate was 69.6% that can be rationalized by high adsorption to sewage sludge and suspended solids [24,25]. However, hydrolysis is not expected to contribute to their removal as its structure lack functional groups that may trigger hydrolysis under environmental condition, also volatilization has negligible effect (low Kc henry law constant). High removal efficiency (74 %) was reported by Fernández-lópez et al. (2016). However, there are other studies that reported low removal efficiency of 40 % in Japan [26] and 17% in Germany [27]. Low removal efficiency of diclofenac could be due to the presence of chlorine group in the molecule decreases biodegradation kinetic constants, kbi. The biodegradation process proceeded through the following series: migration of the organic pollutant from the bulk solution to the microorganism, penetrating (to) microorganism membrane and then enzymatic oxidation for the pollutant. In the case of diclofenac, enzymatic oxidation step is slow as chlorine is electron-withdrawing group and the oxidation process needs oxygen which is electrophilic [26,28–30].

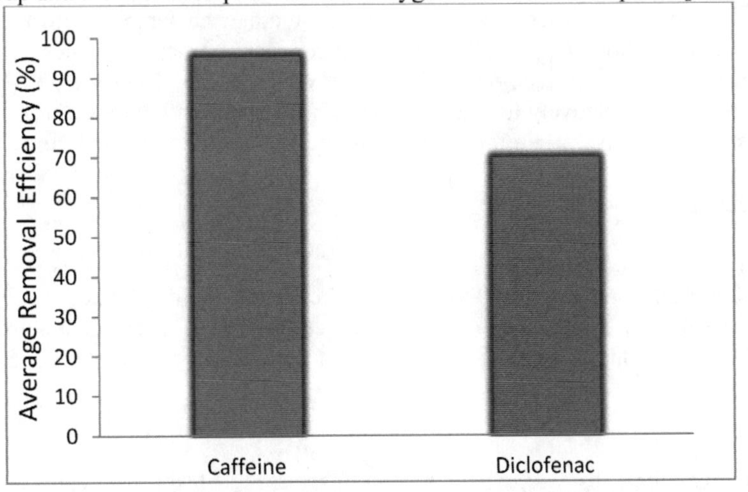

Figure 2. Removal efficiency of WWTP for diclofenac and caffeine.

Caffeine removal from the WWTP was very high that are with large commitment with other results 94.8% [3], 99% [22], 96% [31], >99% [32]. Low removal recorded by Santos et al., 2007 (38 to 86%).Caffeine main removal mechanism is biodegradation because of its slow photo catalysis and its low sorption [8,33].

3.3. Risk assessment

The environmental risk assessment gives us the priorities for the compound that showed a risk for further studying and how to treat it. As shown in table 3, the acute toxicity of diclofenac and caffeine has been calculated for fish, algae and invertebrate and based on the maximum concentration in the effluent of the Waste water treatment plant to present the worthiest case scenario. It was observed that there is only a high risk from diclofenac on fish (RQ>1) and low risk on the other species. Diclofenac high risk was also observed in other studies that reported high diclofenac concentration than our study[3,10]. On the other hand, there is a low risk from caffeine on all the studied species ($0.01 \leq RQ < 0.1$) although caffeine concentration is higher than diclofenac. This observation reflects that the toxicity of the pollutants depends on many factors other than concentration.

Table 3. Acute toxicity of the studied pharmaceuticals.

Compound	Species	EC_{50},LC_{50}(mg/L)	References	PNEC (mg/L)	MEC (mg/L)	RQ
Diclofenac	Algae	16	[34]	0.016	0.00065	0.04
	Fish	0.09	[35]	$9\times10^{\wedge-5}$		**7.22**
	Invertebrates	23	[34]	0.023		0.028
Caffeine	Algae	46	[36]	0.046	0.00232	0.05
	Fish	87.5	[37]	0.0875		0.026
	Invertebrates	182		0.182		0.013

4. Conclusion

Both of caffeine and diclofenac were detected in all of the collected samples. However, caffeine was detected in high concentration (27.85 μg/L– maximum concentration in the influent) and showed high removal efficiency (95.2%), its ecological risk is low for algae, Daphnia and fish. Moreover, diclofenac was detected in relatively low concentration in comparison with caffeine, and was found to have a high risk effect on fish. Therefore, further studies on detection of diclofenac and caffeine in surface and drinking water is essential, also the acute and chronic environmental risk on different organisms is required.

Acknowledgement
This work was financially supported by the by Scientific Research Development Unit (SRDU), Beni-Suef University (BSU). The authors are grateful to Central Lab of Tezmant municipal wastewater treatment plant, Potable water & Sanitation Company in Beni-Suef governorate, Egypt, for helping us in collecting the samples.

References

[1] K'oreje K O, Vergeynst L, Ombaka D, De Wispelaere P, Okoth M, Van Langenhove H and Demeestere K 2016 Occurrence patterns of pharmaceutical residues in wastewater, surface water and groundwater of Nairobi and Kisumu city, Kenya *Chemosphere* **149** 238–44
[2] Crane M, Watts C and Boucard T 2006 Chronic aquatic environmental risks from exposure to human pharmaceuticals *Sci. Total Environ.* **367** 23–41
[3] Papageorgiou M, Kosma C and Lambropoulou D 2016 Seasonal occurrence, removal, mass loading and environmental risk assessment of 55 pharmaceuticals and personal care products in a municipal wastewater treatment plant in Central Greece *Sci. Total Environ.* **543** 547–69
[4] Richardson B J, Lam P K S and Martin M 2005 Emerging chemicals of concern: pharmaceuticals and personal care products (PPCPs) in Asia, with particular reference to Southern China *Mar. Pollut. Bull.* **50** 913–20
[5] Verlicchi P, Aukidy M Al and Zambello E 2012 Occurrence of pharmaceutical compounds in urban wastewater : Removal , mass load and environmental risk after a secondary treatment — A review *Sci. Total Environ.* **429** 123–55
[6] Kümmerer K 2009 Antibiotics in the aquatic environment--a review--part I. *Chemosphere* **75** 417–34
[7] Ternes T A 1998 Occurrence of drugs in German sewage treatment plants and rivers *Water Res.* **32** 3245–60
[8] Buerge I J, Poiger T, Müller M D and Buser H-R 2003 Caffeine, an antropogenic marker for wastewater contamination of surface waters *Environ. Sci. Technol.* **37** 691
[9] Carmona E, Andreu V and Picó Y 2014 Occurrence of acidic pharmaceuticals and personal care products in Tuna River Basin : From waste to drinking water *Sci. Total Environ.* **484** 53–63
[10] Kosma C I, Lambropoulou D A and Albanis T A 2014 Investigation of PPCPs in wastewater treatment plants in Greece: occurrence, removal and environmental risk assessment *Sci. Total Environ.* **466** 421–38
[11] Vieno N, Tuhkanen T and Kronberg L 2007 Elimination of pharmaceuticals in sewage treatment plants in Finland *Water Res.* **41** 1001–12
[12] Collier A C 2007 Pharmaceutical contaminants in potable water: potential concerns for pregnant women and children *Ecohealth* **4** 164–71
[13] Oaks J L, Gilbert M, Virani M Z, Watson R T, Meteyer C U, Rideout B A, Shivaprasad H L, Ahmed S, Chaudhry M J I and Arshad M 2004 Diclofenac residues as the cause of vulture population decline in Pakistan *Nature* **427** 630
[14] Hartmann J, Bartels P, Mau U, Witter M, Tümpling W v., Hofmann J and Nietzschmann E 2008 Degradation of the drug diclofenac in water by sonolysis in presence of catalysts *Chemosphere* **70** 453–61

[15] Martínez C, Canle M L, Fernández M I, Santaballa J A and Faria J 2011 Applied Catalysis B : Environmental Aqueous degradation of diclofenac by heterogeneous photocatalysis using nanostructured materials *Appl. Catal. B Environ.* **107** 110–8

[16] R. T, H. C, A. H, R. E and H.-R. K 2004 Toxic effects of the non-steroidal anti-inflammatory drug diclofenac: Part II. Cytological effects in liver, kidney, gills and intestine of rainbow trout (Oncorhynchus mykiss) *Aquat. Toxicol.* **68** 151–66

[17] Schwaiger J, Ferling H, Mallow U, Wintermayrb H and Negelea R D 2004 Toxic effects of the non-steroidal antiinflammatory drug diclofenac: part I: histopathological alterations and bioaccumulation in rainbow trout *Aquat Toxicol* **68** 141

[18] Santos L H M L M, Gros M, Rodriguez-mozaz S, Delerue-matos C, Pena A, Barceló D and Montenegro M C B S M 2013 Science of the Total Environment Contribution of hospital ef fl uents to the load of pharmaceuticals in urban wastewaters : Identi fi cation of ecologically relevant pharmaceuticals *Sci. Total Environ.* **461–462** 302–16

[19] Kosma C I, Lambropoulou D A and Albanis T A 2010 Occurrence and removal of pharmaceutically active compounds in wastewaters in Greece *HazWasManagement2010* **179** 1–4

[20] European Commission - Joint Research Centre - Institute for Health and Consumer Protection 2003 Technical Guidance Document on Risk Assessment *Inst. Heal. Consum. Prot. Eur. Chem. Bur. Part II. Available online http//echa. Eur. eu/documents/10162/16960216/tgdpart2_2ed_en. pdf* 398

[21] Tewari S, Jindal R, Kho Y L, Eo S and Choi K 2013 Major pharmaceutical residues in wastewater treatment plants and receiving waters in Bangkok , Thailand , and associated ecological *Chemosphere* **91** 697–704

[22] Gómez M J, Martínez Bueno M J, Lacorte S, Fernández-Alba A R and Agüera A 2007 Pilot survey monitoring pharmaceuticals and related compounds in a sewage treatment plant located on the Mediterranean coast *Chemosphere* **66** 993–1002

[23] Zhou, Haidong, Chunying Wu, Xia Huang, Mijun Gao, Xianghua Wen, Hiroshi Tsuno and H T 2010 Occurrence of Selected Pharmaceuticals and Caffeine in Sewage Treatment Plants and Receiving Rive ... *Science (80-.).* **82** 2239–48

[24] Sipma J, Osuna B, Collado N, Monclús H, Ferrero G, Comas J and Rodriguez-Roda I 2009 Author's personal copy Comparison of removal of pharmaceuticals in MBR and activated sludge systems ☆ *Desalination* **250** 653–9

[25] Ternes T A, Herrmann N, Bonerz M, Knacker T, Siegrist H and Joss A 2004 A rapid method to measure the solid-water distribution coefficient (K-d) for pharmaceuticals and musk fragrances in sewage sludge *Water Res.* **38** 4075–84

[26] Fernández-López C, Guillén-Navarro J M, Padilla J J and Parsons J R 2016 Comparison of the removal efficiencies of selected pharmaceuticals in wastewater treatment plants in the region of Murcia, Spain *Ecol. Eng.* **95** 811–6

[27] Kimura K, Hara H and Watanabe Y 2007 Elimination of selected acidic pharmaceuticals from municipal wastewaters by activated sludge systems and membrane bioreactors *Environ. Sci. Technol.* **41** 3708–14

[28] Heberer T, Reddersen K and Mechlinski A 2002 From municipal sewage to drinking water: fate and removal of pharmaceutical residues in the aquatic environment in urban areas. *Water Sci. Technol.* **46** 81–8

[29] Jelic A, Gros M, Ginebreda A, Cespedes-Sánchez R, Ventura F, Petrovic M and Barcelo D 2010 Title: Occurrence, partition and removal of pharmaceuticals in sewage water and sludge during wastewater treatment partition and removal of pharmaceuticals in sewage water and sludge during wastewater treatment *Water Res.* **45** 1165–76

[30] Cirja M and Ivashechkin Æ P 2007 Factors affecting the removal of organic micropollutants from wastewater in conventional treatment plants (CTP) and membrane bioreactors (MBR) Factors affecting the removal of organic micropollutants from wastewater in conventional treatment plants (*Rev. Environ. Sci. Bio/Technology* **7** 61–78

[31] Andreozzi R, Cesaro R, Marotta R and Pirozzi F 2006 Evaluation of biodegradation kinetic

constants for aromatic compounds by means of aerobic batch experiments *Chemosphere* **62** 1431–6

[32] Heberer T 2016 Tracking persistent pharmaceutical residues from municipal sewage to drinking water . J sewage to drinking water *J. Hydrol.* **266** 175–89

[33] Ternes T A, Bonerz M and Schimdt T 2001 Determination of neutral pharmaceuticals in waste waters and rivers by liquid chromatography electrospray tandem mass spectrometry *J. Chromatogr. A* **938** 175–85

[34] Grung M, Källqvist T, Sakshaug S, Skurtveit S and Thomas K V. 2008 Environmental assessment of Norwegian priority pharmaceuticals based on the EMEA guideline *Ecotoxicol. Environ. Saf.* **71** 328–40

[35] Dietrich D R and Prietz A 1999 Fish embryotoxicity and teratogenicity of pharmaceuticals, detergents and pesticides regularly detected in sewage treatment plant effluents and surface waters *Toxicologist* **48** 151

[36] Sanderson H, Johnson D ., Wilson C ., Brain R . and Solomon K . 2003 Probabilistic hazard assesment of environmentally occuring pharmaceuticals toxicity to fish, daphnids,and algae by ECOSAR screening *Toxicol. Lett.* **144** 383–95

[37] Fernández C, González-Doncel M, Pro J, Carbonell G and Tarazona J V. 2010 Occurrence of pharmaceutically active compounds in surface waters of the henares-jarama-tajo river system (madrid, spain) and a potential risk characterization *Sci Total Env.* **408** 543–51

IOP Conf. Series: Materials Science and Engineering **464** (2018) 012002 doi:10.1088/1757-899X/464/1/012002

A promising fuel cell catalyst using non-precious metal oxide

Hamdy F M Mohamed[1,2*], E E Abdel-Hady[1], M F M Hmamm[2], M Ibrahim[1], H Ahmed[1], M Mondy[1] and H Yehia[1]

[1] Physics Department, Faculty of Science, Minia University, P.O. Box 61519 Minia, Egypt.
[2] Renewable Energy Science & Engineering Department, Faculty of Postgraduate Studies for Advanced Science (PSAS), Beni-Suef University, P.O. Box 62511 Beni-Suef, Egypt
* Corresponding author, e-mail: hamdyfm@gmail.com&hamdy.farghal@mu.edu.eg

Abstract. Fuel cell has two essential problems, its cost and the durability, which hinder its commercialization. Platinum is the ideal catalyst that has high activity, stability and selectivity but has high cost. An attempt has been done to find a cheaper catalyst instead of platinum. Zinc oxide nanoparticles were synthesized via sol-gel method using zinc acetate and citric acid in basic media with different calcination temperatures (420, 520 and 620 ^{o}C). From X-ray diffraction (XRD) patterns, the calculated particles size is 7.7, 15.6 and 19.3 nm as the calcination temperature of 420, 520 and 620°C, respectively, indicating that the particles size increases with increasing the calcination temperature. Different concentrations (5 and 10 wt.%) of ZnO nanoparticles with 10 wt.% polyvinyl alcohol (PVA) were prepared and calcinated at 750 ^{o}C to get carbon/ZnO as a catalyst for fuel cell applications. A carbon core-shell surrounding by mono-disperse ZnO nanoparticles with large surface area that required for the new catalyst with believable morphology was shown by transmission electron microscope (TEM). Also, XRD presents high purity of the new composite with uniformly distinguishable peaks. Fourier transformation infrared (FTIR) spectroscopy shows the change in the carbon/ZnO nanoparticles spectra due to ZnO characteristic vibration band at 440-460 cm^{-1}. Cyclic voltammetry (CV) exhibits a good promising catalytic activity and current density with oxidation behaviour is reported. Finally, ZnO used to enhance carbon electrochemically performance as a result of a novel non-precious catalyst.

1. Introduction

Energy is a key element of the interaction between nature and society and affects the economic development as well as the future of human existence. Concerns associated with human health and environmental risks have arisen rapidly over the last decades. Many of these issues are related to the production, transformation and use of energy, for example acid rain, ozone depletion and global climate change. Scientists focus their efforts into developing energy production processes that will be beneficial towards a sustainable development and simultaneously will satisfy society's energy needs. Meeting the present needs without compromising the future should be achieved by using alternative sources of energy like the hydrodynamic, solar, wind, geothermal, etc. Development of new energy system is necessary for sustaining growing economies, disengaged from oil. Fuel cells are one of the key future technologies similar to batteries with the exception of the continuous replenishment of the fuel. They promise to reduce the ecological impact by creating environmentally friendly by products while using renewable fuel sources. They can be employed in a variety of applications, due to their flexible power output and ranging from a few watts to megawatts. Understanding fuel cells are thus pertinent to the advancement and sustainability of our energy demanding lifestyles.

The key components of the polymer electrolyte membrane fuel cells (PEMFCs), as shown in Figure 1, are: (a) the anode that consists of a support, a diffusion layer and the catalyst where the oxidation reaction produces electrons that flow through the external load towards the cathode; (b) the electrolyte membrane whose aqueous pockets provide a path for the protons to diffuse and reach the cathode side; and (c) the cathode where the protons react with oxygen to produce water on the hosted catalyst/diffusion layer. Thus, the transport of reactants towards the reaction sites is crucial and can result in significant efficiencies [1]. In PEMFC, hydrogen (H_2) is fed continuously to the anode in which it oxidized and oxygen (O_2) reduced at cathode. Electrolyte membrane permits protons resulting from H_2 oxidation to pass to other side. As electrolyte membrane is electrically insulator so electrons released from broken H_2 bond could pass only through external circuit generating current. Water is produced as hydrogen ions recombine with reduced oxygen ions [2,3] as;

$$H_2 \rightarrow 2H^+ + 2e^- \quad (1)$$

$$\tfrac{1}{2}O_2 + 2H^+ + 2e^- \rightarrow H_2O \quad (2)$$

$$H_2 + \tfrac{1}{2}O_2 \rightarrow H_2O \quad (3)$$

Catalyst at both electrodes used for speeding oxidation and reduction process so there are different requirements should be existed for good performance catalyst as activity, selectivity, stability and poising resistance [4]. Platinum (Pt) is the best element that meets the requirements. The biggest barrier facing fuel cells to commercialize is the high price of Pt ratio that used. Considerable attentions for replacing Pt are made by alloying [5,6], layering [7,8], core-shell [9–16] and free Pt techniques. The use of transition-metal nanoparticles in catalysis is crucial as they mimic metal surface activation, selectivity and efficiency to heterogeneous catalysis [17,18]. Among transition-metal nanoparticles, ZnO has been of considerable interest because of the role of ZnO in solar cells, catalysts, antibacterial materials, gas sensors, luminescent materials, and photocatalyst [19]. Nano-ZnO, as heterogeneous catalyst, has received great attention because it is inexpensive, nontoxic catalyst. It has also environmental advantages, which is, minimum execution time, low corrosion, waste minimization, recycling of the catalyst, easy transport, and disposal of the catalyst. So, the aim of this work is to prepare and characterize ZnO nanoparticles with carbon support which work as a free Pt catalyst for fuel cell applications. Characterizations of the samples were done using x-ray diffraction (XRD), Fourier transformation infrared spectroscopy (FTIR), transmission electron microscopy (TEM) and cyclic voltammetry (CV).

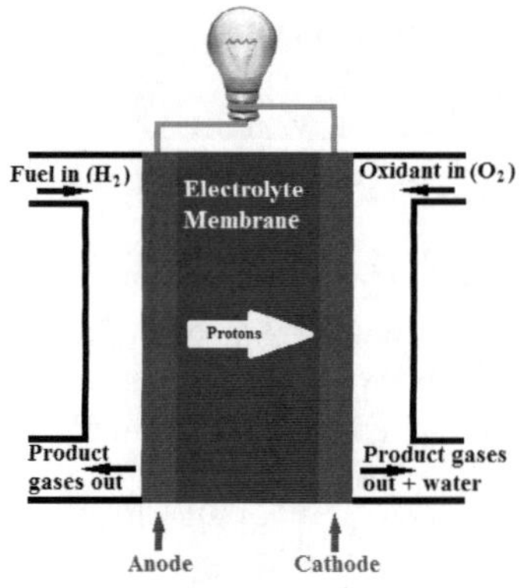

Figure 1. The polymer electrolyte membrane fuel cell (PEMFC).

2. Experimental setup

Zinc acetate dehydrate $Zn(O_2CCH_3)_2$ with purity more than 97% was purchased from Alfa Aeser, Germany, as a zinc source. A citric acid $C_6H_8O_7$ was purchased from Adwic, Egypt, and was used to react with the acetate to separate the zinc acetate to its parts. Zinc oxide nanoparticles were prepared using sol-gel method. This preparation is one of the simplest, environmental safe and lowest-cost technique for preparing pure transition metal oxides with relatively high surface area at low temperature. Adding ammonia solution was necessary for maintain the reaction pH at 8.5, prevent the solution from perception and get mono dispersal particles. Reagents and solvents were obtained from commercial sources and used without further purification. Three different samples were prepared at different calcination temperatures (420, 520 and 620 °C) with constant time (5h) as shown in Figure 2.

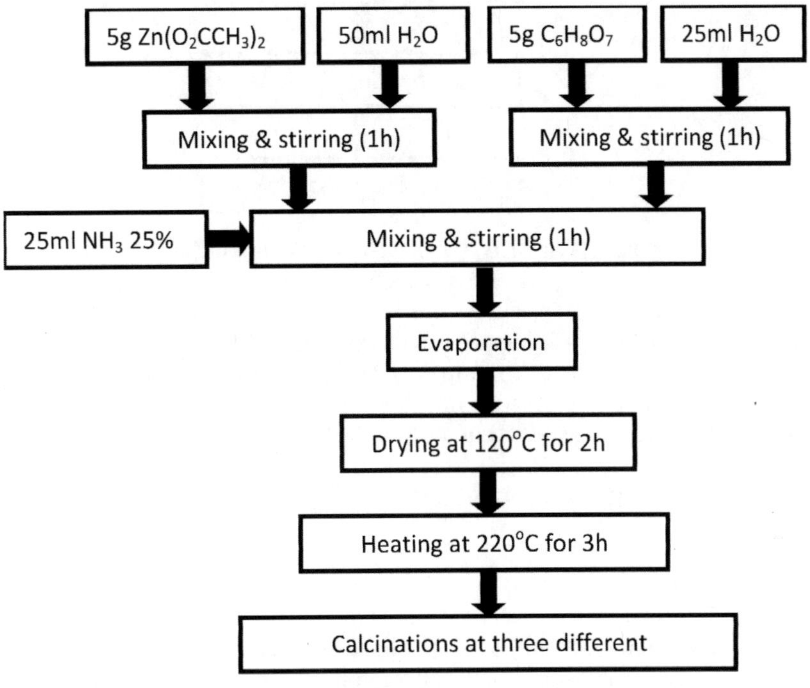

Figure 2. Synthesis of ZnO nanoparticles.

Polyvinyl alcohol (PVA) extra purity has purchased from Alpha Chemika with molecular weight 125000, which has low electrical conductivity, has been used as a carbon source. The mixture solution was prepared by dissolving PVA, 98-99% hydrolyzed in 20 ml of deionized water (10 wt.%). It was heated at 60 °C until the solution became clear (5–7h). Zinc oxide nanoparticles with two different concentrations (5 and 10 wt.%) were being dissolved in 5 ml of deionized water then added to the PVA solution and stirring the mixture for 24h until complete dissolving. Figure 3 shows the flowchart for the sample preparation.

X-ray diffraction (XRD) was measured for the present samples by Ultima IV (Rigaku Corporation) with a 285 mm goniometer. A parallel monochromatic x-ray beam acquired by reflection of CuKα radiation (λ = 1.542 Å) from a fine focus x-ray tube worked at 40 kV and 40 mA. Scattering profiles were recorded with a scintillation one-dimension position-sensitive detector over a range of $5^o < 2\theta < 70^o$ with a step of 0.01^o. The chemical structure of the samples under study turned into examined through FTIR using VERTEX 70 spectrometer, Bruker, USA, from wavenumber 4000 to 400 cm^{-1} with resolution 0.9 cm^{-1} at room temperature (25 °C).

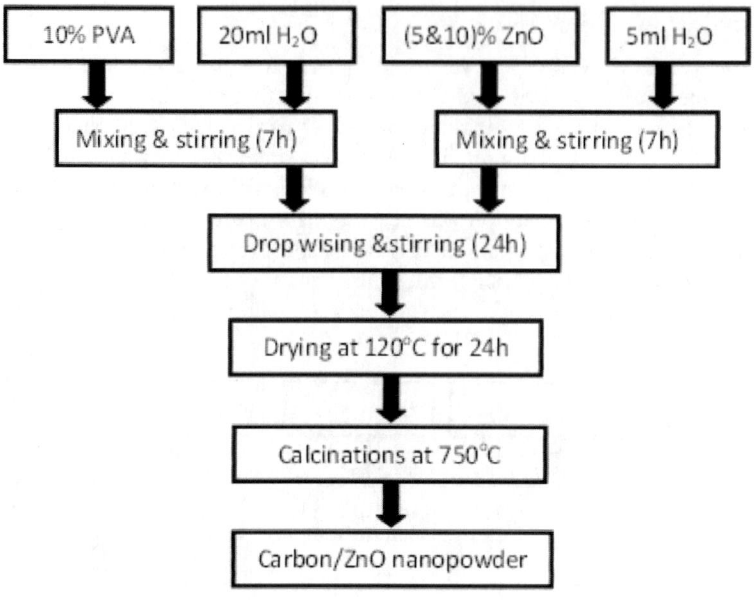

Figure 3. Synthesis of nanopowder of carbon/ZnO catalyst.

For electrochemical kinetic measurements, cyclic voltammetry (Autolab 302n potentiostat) at the National Research Center, Egypt, was used in the range of potential from -1 to -0.02 V with scan rates 50 and 100 mV/s and two different types of electrolytes (1M KOH and mixture of 1M KOH & 2M Methanol). Transmission electron microscopy (TEM) for carbon/ZnO nanoparticles 5 and 10 wt.% was carried out using JEOL JEM2100 made in USA for studying the surface and morphology. A few droplets of powder, ultrasonically dispersed in alcohol, had been put on standard microscope grid for the TEM work.

3. Results and Discussion

3.1. X-ray diffraction

X-ray diffraction (XRD) was used as crystal structure analysis for the three samples of ZnO nanoparticles calcinated at 420, 520, and 620 °C. Figure 4 illustrates the XRD pattern for the three different samples with very sharp peaks and narrow full width at half maximum (FWHM) which indicate the crystallinity in nature with clear diffraction peaks for wurtzite phase as mentioned in the standard data [19,20]. It is clear from the figure that, with increasing the calcination temperatures, the intensity of the peaks increases and also the FWHM decreases, that strongly suggesting good crystallinity of the nanomaterials. The peaks at 2θ = 31.67°, 34.31°, 36.14°, 47.40°, 56.52°, 62.73°, 66.28°, 67.91°, 69.03°, and 72.48° were assigned to (100), (002), (101), (102), (110), (103), (200), (112), (201), and (004), respectively of ZnO nanoparticles, indicating that the samples were polycrystalline wurtzite structure (Zincite, JCPDS 5-0664) [20–22]. For the three samples, no corresponding peaks for reactants remained were observed indicating that the prepared samples have a high purity. Sherrer's equation (equation (4)) could be used to determine the particle size as:

$$L_{hKL} = \frac{K\lambda}{B\cos(\theta)}. \quad (4)$$

Where L_{hkL} is the mean size of the particle, K is a dimensionless shape factor with a value close to unity, λ is the x-ray wavelength, β is the line broadening at the full width at half maximum (FWHM) in radians, and θ is the Bragg angle (in degrees). Table 1 lists the average particle size at different calcination temperatures for ZnO nanoparticles. It is clear from the table that, there is a correlation between varying the calcination temperature and the average particles size. Increasing in calcination temperature leads to increasing the grain size as the crystal structure rearranged, crystals agglomerate with each other, and removal of traces. Thus, in order to get smaller particles lower calcinations temperature is favourable. By comparing the XRD pattern of the three samples it has been concluded that samples calcinated at 520 and 620 °C gives high intensity fine peaks, which can be used for further preparation and characterization.

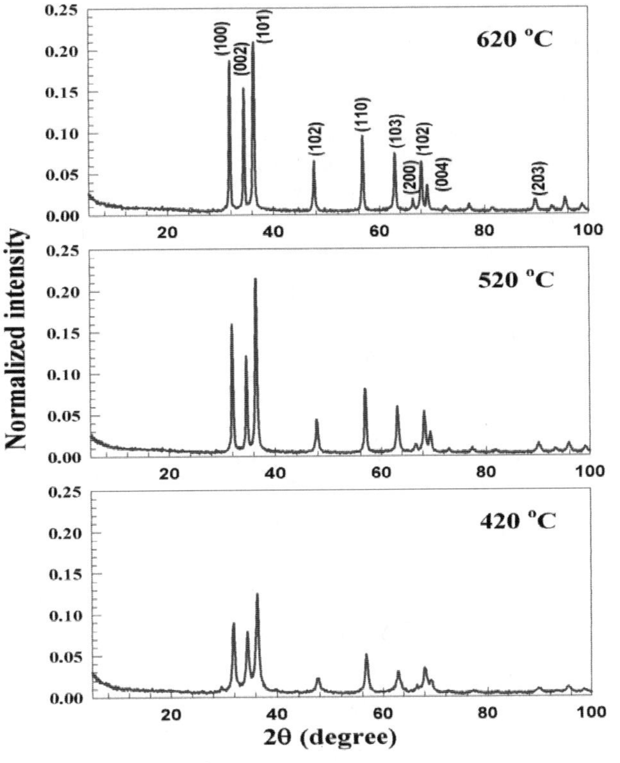

Figure 4. The XRD for the three ZnO nanoparticles calcinated at different temperatures.

The XRD for the two different concentrations of carbon/ZnO nanoparticles (5 and 10 wt.%) are similar and the high concentration is presented in Figure 5. It is noticed that, there is no significant difference between this XRD pattern and the other one in Figure 4 except the appearance of carbon hump at $2\theta \sim 22^{o}$ similar to the data of Biswas *et al.* [23] and also decreasing the intensity of some peaks. This may be due to physical interaction between carbon and ZnO nanoparticles.

Table 1. Average particle sizes at different calcination temperature for ZnO nanoparticles.

Calcination temperature (°C)	Average particles size (nm)
420	7.7
520	15.6
620	19.3

3.2. Fourier transformation infrared spectroscopy (FTIR)

FTIR used to study the catalyst's molecular structure as shown in Figure 6 for carbon/ZnO nanoparticles 5 and 10 wt.%. The figure shows a series of absorption peaks from 4000 to 400 cm^{-1}, which are corresponding to the carboxylate and hydroxyl impurities in the samples. For both samples, band at 3500 cm^{-1} is caused by stretching of hydroxyl group (O-H) and from 2700 to 3000 cm^{-1} band is corresponding to stretching vibration of C-H of alkane groups. The peaks at 1600 and 1100 cm^{-1} are the results of stretching asymmetrically and symmetrically of the zinc carboxylate. Also, the characteristic bands at 459 and 445 cm^{-1} for carbon/ZnO nanoparticles of 5 and 10 wt.%, respectively, could be due to the vibration of the ZnO nanoparticles [24]. According to Verges [25], the single band centered at 460 cm^{-1} must be assigned to spherical particles, but when splitting occurs, prolative or platelike shapes are present. As the concentration of carbon (PVA) increases, the content of the carboxylate (COO^-) and hydroxyl (^-OH) groups in the samples decreased. The carboxylate probably comes from both of reactive carbon containing acetate species during synthesis and the hydroxyl results from the hygroscopic nature of ZnO. As a result, FTIR spectra suggest that some impurity exists near ZnO surface. This may be due to low calcination temperature and short time which they were not enough more to execute it as reported in the literature [26,27].

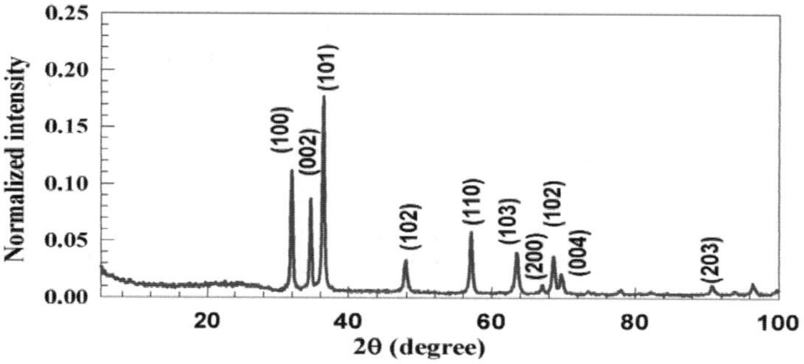

Figure 5. The XRD for carbon/ZnO (10 wt.%) nanoparticles.

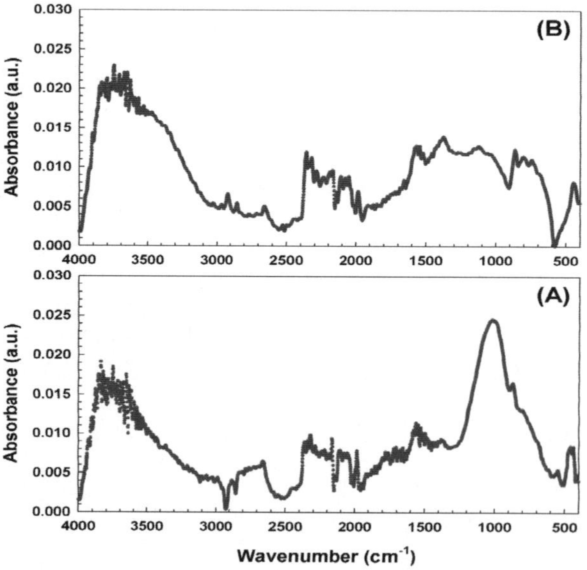

Figure 6. The FTIR spectra of carbon/ZnO nanoparticles (A) 5 wt.% and (B) 10 wt.%.

3.3. Transmission Electron Microscopy (TEM)

TEM images for the carbon/ZnO nanoparticles 5 and 10 wt.% are shown in Figure 7 and 8, respectively. From Figure 7, it is observed that there are nanoparticles of ZnO in the surface of the carbon with different morphology and randomly distribution. The ZnO nanoparticles are surrounding by carbon which provide more surface area. It well known that, larger active surface area is recommended for the catalyst requirements [28]. The average size of ZnO nanoparticles is 19.42 nm which is in a good agreement with the XRD data [Table 1]. For high concentration of ZnO nanoparticles [Figure 8], the ZnO nanoparticles are widely distributed over a large range of carbon. In addition, the nanoparticles of ZnO had a better morphology (may be arranged as hexagonal crystals) than that in low concentration [Figure 7]. Also, there is a very good distribution of ZnO nanoparticles over a wide range of carbon compared to low concentration sample which support more surface area. Increasing the concentration of the ZnO nanoparticles affects the aggregations of nanoparticles over the carbon that make a core-shell of carbon surrounded by a high catalytic ZnO nanoparticles with large active surface area. From Figure 8, the average size of the carbon core-shell that surrounding by ZnO nanoparticles is found to be about 40 nm.

3.4. *Cyclic voltammetry (CV)*

Carbon/ZnO nanostructures (5 & 10 wt.%) were activated by applying cyclic voltages in 1M KOH and mixture of 1M KOH and 2M methanol electrolytes. The cyclic voltammetric behaviors of prepared samples are shown in Figure 9 with scan rate of 100 mV/s. Also, the cyclic voltammetric behaviors of standard Pt/C which is one of the most efficient commercial catalysts that has high activity is included. Carbon has low catalytic activity [29] that enhanced by adding ZnO nanoparticles with two different concentrations (5&10 wt.%). The electrocatalytic properties of carbon/ZnO powder were evaluated and compared with the commercial Pt/C. Table 2 lists the current density and applied voltage with 50 and 100 mV/s scan rate for the carbon/ZnO nanoparticles 5& 10 wt.% and also for Pt/C.

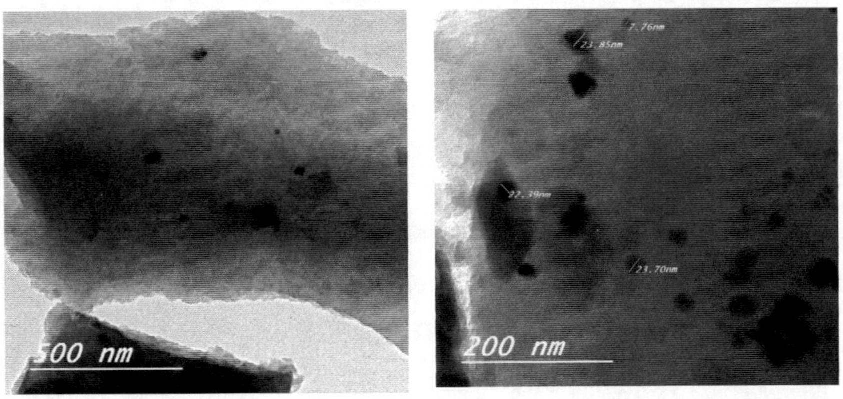

Figure 7. TEM images for the carbon/ZnO nanoparticles (5 wt.%) calcinated at 750 °C for 2h.

Figure 8. TEM images for the carbon/ZnO nanoparticles (10 wt.%) calcinated at 750 °C for 2h.

Carbon/ZnO 10 wt.% nanoparticles showed good oxidation behavior in KOH+methanol environment as the current density peak reaches 0.35 mA/cm^2 with scan rate 100 mV/s. On the other hand, the current density for carbon/ZnO (5 wt.%) was found to be 0.17 mA/cm^2. Increasing nanoparticles concentration increases the adsorption of OH in the electrolyte as more surface area was introduced for oxidation of methanol. Methanol oxidation peaks in anodic direction are reported with different intensities for carbon/ZnO nanoparticles (5&10 wt.%). Pt/C showed less current density in KOH reaches at maximum 0.55 mA/cm^2 while the highest current density of 5 wt.% sample is 0.06 mA/cm^2 at the same condition. For 10 wt.% sample, the oxidation peak height is almost three times as that in the 5 wt.% sample and nearly 33% of that for commercial Pt/C [Table 2]. Generally, it is understood that the anodic peak in the reverse scan to be linked to the removal of incompletely oxidized carbonaceous species accumulated on the catalyst surface during the forward anodic scan [30,31]. The oxidation peaks of KOH at anodic direction could be observed in all formulations. The obtained results of carbon/ZnO (10 wt.%) are comparative to commercial Pt/C activity which could be enhanced using carbon core-shell surrounded by more concentrations of ZnO nanoparticles with higher calcination temperatures.

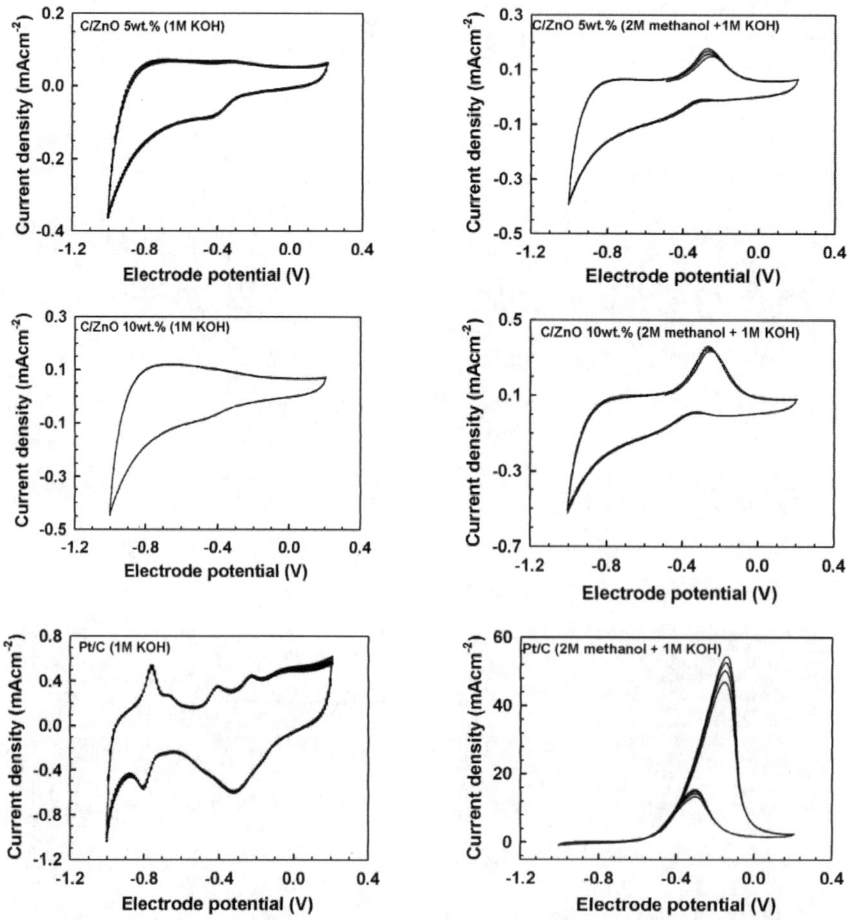

Figure 9. Consecutive CV of carbon/ZnO nanoparticles (5 & 10 wt.%) and Pt/carbon at scan rate of 100 mVs^{-1}.

Table 2. Current density and applied voltage for the carbon/ZnO nanoparticles 5 and 10 wt.% with different scan rates and different electrolyte. The data for Pt/C is also included for comparison.

Conditions/Samples		Pt/C		Carbon/ZnO (5 wt.%)		Carbon/ZnO (10 wt.%)	
Electrolyte	Scan rate (mV/s)	Current Density (mA/cm^2)	Applied voltage (V)	Current Density (mA/cm^2)	Applied voltage (V)	Current Density (mA/cm^2)	Applied voltage (V)
1M KOH	50	0.36	0.25	0.04	- 0.7	0.12	0.76
	100	0.55	0.28	0.06	- 0.75	0.14	0.76
2M Methanol +1M KOH	50	51	-0.2	0.14	-0.2	0.27	-0.25
	100	55	-0.2	0.17	-0.2	0.35	-0.28

4. Conclusion

ZnO nanoparticles with large surface area by sol-gel method were successfully synthesised with undertaken different calcination temperatures. XRD results for ZnO nanoparticles showed that the particles size increased directly with increasing calcination temperatures as the crystal structure rearranged, crystals agglomerated with each other, and traces were removed. The optimized temperature for preparing carbon/ZnO catalyst was 620 °C. According to XRD pattern and FTIR spectrum for the new catalyst, ZnO nanoparticles were adsorbed over carbon surface as physical interaction could be noticed. TEM images determined that carbon/ZnO has large surface area that is required for better catalytic activity. The core-shell of carbon surrounded by ZnO that successfully prepared provides high stability for the new catalyst. CV results proved that carbon active surface area and electrocatalytic performance enhanced by adding ZnO. Electrocatalytic properties of carbon increased in oxidation cycle by increasing the ZnO concentrations. The new catalyst using carbon support with different concentrations of ZnO nanoparticles improved the catalytic activity, thermal stability and surface corrosion for carbon. Carbon/ZnO was compared with commercial Pt/C and found that the sample of 10 wt. % efficiency reaches 33% of commercial Pt/C. Finally, carbon/ZnO catalyst after particular improvements can be used as a new generation of fuel cell operating by non-platinum catalyst.

Acknowledgement
The authors would like to thank the Academy of Scientific Research and Technology in Egypt (ASRT) for their financial support.

References

[1] O'Hayre R, Cha S W, Colella W and Prinz F B 2006 Fuel cell fundamentals
[2] Kordesch K S 1996 *Fuel cells and their applications* vol 117 (VCh Weinheim)
[3] Hoogers G, Chen E, Thompsett D, Hogarth M, Stone R and Hart D 2003 *Fuel Cell Technology Handbook* (CRC press)
[4] Holton O T and Stevenson J W 2013 The role of platinum in proton exchange membrane fuel cells *Platin. Met. Rev.* **57** 259–71
[5] Greeley J, Stephens I E L, Bondarenko A S, Johansson T P, Hansen H A, Jaramillo T F, Rossmeisl J, Chorkendorff I and Nørskov J K 2009 Alloys of platinum and early transition metals as oxygen reduction electrocatalysts *Nat. Chem.* **1** 552–6
[6] Stamenkovic V, Mun B S, Mayrhofer K J J, Ross P N, Markovic N M, Rossmeisl J, Greeley J and Nørskov J K 2006 Changing the Activity of Electrocatalysts for Oxygen Reduction by Tuning the Surface Electronic Structure *Angew. Chemie* **118** 2963–7
[7] Endoh E, Terazono S, Widjaja H and Takimoto Y 2004 Degradation Study of MEA for PEMFCs under Low Humidity Conditions *Electrochem. Solid-State Lett.* **7** A209
[8] Kinumoto T, Inaba M, Nakayama Y, Ogata K, Umebayashi R, Tasaka A, Iriyama Y, Abe T

and Ogumi Z 2006 Durability of perfluorinated ionomer membrane against hydrogen peroxide *J. Power Sources* **158** 1222–8

[9] Stephens I E L, Bondarenko A S, Perez-Alonso F J, Calle-Vallejo F, Bech L, Johansson T P, Jepsen A K, Frydendal R, Knudsen B P, Rossmeisl J and Chorkendorff I 2011 Tuning the Activity of Pt(111) for Oxygen Electroreduction by Subsurface Alloying *J. Am. Chem. Soc.* **133** 5485–91

[10] Zhao D and Xu B-Q 2006 Enhancement of Pt Utilization in Electrocatalysts by Using Gold Nanoparticles *Angew. Chemie Int. Ed.* **45** 4955–9

[11] Zhao D and Xu B-Q 2006 Platinum covering of gold nanoparticles for utilization enhancement of Pt in electrocatalysts *Phys. Chem. Chem. Phys.* **8** 5106–14

[12] Wang G, Huang B, Xiao L, Ren Z, Chen H, Wang D, Abruña H D, Lu J and Zhuang L 2014 Pt Skin on AuCu Intermetallic Substrate: A Strategy to Maximize Pt Utilization for Fuel Cells *J. Am. Chem. Soc.* **136** 9643–9

[13] Wang Y and Toshima N 1997 Preparation of Pd–Pt Bimetallic Colloids with Controllable Core/Shell Structures *J. Phys. Chem. B* **101** 5301–6

[14] Lim Y, Kim S K, Lee S-C, Choi J, Nahm K S, Yoo S J and Kim P 2014 One-step synthesis of carbon-supported Pd@Pt/C core–shell nanoparticles as oxygen reduction electrocatalysts and their enhanced activity and stability *Nanoscale* **6** 4038–42

[15] Zhou S, Varughese B, Eichhorn B, Jackson G and McIlwrath K 2005 Pt–Cu Core–Shell and Alloy Nanoparticles for Heterogeneous NOx Reduction: Anomalous Stability and Reactivity of a Core–Shell Nanostructure *Angew. Chemie Int. Ed.* **44** 4539–43

[16] Chen Y, Liang Z, Yang F, Liu Y and Chen S 2011 Ni–Pt Core–Shell Nanoparticles as Oxygen Reduction Electrocatalysts: Effect of Pt Shell Coverage *J. Phys. Chem. C* **115** 24073–9

[17] Durand J, Teuma E and Gómez M 2008 An Overview of Palladium Nanocatalysts: Surface and Molecular Reactivity *Eur. J. Inorg. Chem.* **2008** 3577–86

[18] Kung H H 1989 *Transition metal oxides: surface chemistry and catalysis* vol 45 (Elsevier)

[19] P N D and G 2010 ZnO nanoparticles and their applications-new achievements *Proceedings of the 2ndInternational Conference on Nanotechnology (NANOCON '10)* (Olomouc, Europe)

[20] Kulkarni S S, Sawarkar Mahavidyalaya S and Shirsat M D 2015 Optical and Structural Properties of Zinc Oxide Nanoparticles *Int. J. Adv. Res. Phys. Sci.* **2** 14–8

[21] Shalaby A, Bachvarova-Nedelcheva A, Iordanova R and Dimitriev Y 2013 A study of the effect of citric acid on the crystallinity of ZnO/TiO2 nanopowders *J. Chem. Technol. Metall.* **48** 585–90

[22] Bala N, Saha S, Chakraborty M, Maiti M, Das S, Basu R and Nandy P 2015 Green synthesis of zinc oxide nanoparticles using Hibiscus subdariffa leaf extract: effect of temperature on synthesis, anti-bacterial activity and anti-diabetic activity *RSC Adv.* **5** 4993–5003

[23] Biswas C, Bhattacharyya A, Chandra Das G, Choudhuri M G and Dey R 2016 A Novel Devolatilization Technique of Pre-reduction of Iron Ore Using Lean Grade Coal *BHM Berg- und Hüttenmännische Monatshefte* **161** 95–101

[24] Handore K, Bhavsar S, Horne A, Chhattise P, Mohite K, Ambekar J, Pande N and Chabukswar V 2014 Novel green route of synthesis of ZnO nanoparticles by using natural biodegradable polymer and its application as a catalyst for oxidation of aldehydes *J. Macromol. Sci. Part A* **51** 941–7

[25] Andrés-Vergés M and Martínez-Gallego M 1992 Spherical and rod-like zinc oxide microcrystals: morphological characterization and microstructural evolution with temperature *J. Mater. Sci.* **27** 3756–62

[26] Stoyanova A, Hitkova H, Bachvarova-Nedelcheva A, Iordanova R, Ivanova N and Sredkova M 2013 SYNTHESIS AND ANTIBACTERIAL ACTIVITY OF TiO 2 /ZnO NANOCOMPOSITES PREPARED VIA NONHYDROLYTIC ROUTE *J. Chem. Technol. Metall.* **48** 154–61

[27] Xiong G, Pal U, Serrano J G, Ucer K B and Williams R T 2006 Photoluminesence and FTIR study of ZnO nanoparticles: the impurity and defect perspective *Phys. Status Solidi* **3** 3577–81

[28] Dirote E V 2006 *Nanotechnology at the Leading Edge* (Nova Publishers)

[29] Trimble S W 2007 *Encyclopedia of Water Science, Second Edition - Two Volume Set (Print Version)* (CRC press)

[30] Sun S, Zhang G, Gauquelin N, Chen N, Zhou J, Yang S, Chen W, Meng X, Geng D and Banis M N 2013 Single-atom catalysis using Pt/graphene achieved through atomic layer deposition *Sci. Rep.* **3** 1775

[31] Mancharan R and Goodenough J B 1992 Methanol oxidation in acid on ordered NiTi *J. Mater. Chem.* **2** 875–87

Water quality assessment of Qarun Lake and heavy metals decontamination from its drains using nanocomposites

M E Goher[1], W M A El-Rouby[2], S I El-Dek[2*], S M El-Sayed[1] and S G Noaemy[1,2]

[1] Chemistry Lab., Fresh water and Lakes division, NIOF, National Institute of Oceanography and Fisheries, Egypt.
[2] Materials Science and Nanotechnology Department, Faculty of Postgraduate Studies for Advanced Sciences, (PSAS), Beni-Suef University, Beni-Suef, Egypt.

* Corresponding author: e-mail:didi5550000@gmail.com, samaa@psas.bsu.edu.eg (S.I.ELDek)

Abstract. Qarun Lake is one of the most important lakes in Egypt; it is the shrunken remnant of Moeris Lake. It is a closed ecosystem represented as a tank for discharged agricultural wastewater of El-Fayoum province. To assess the quality of the lake water for fishing and aquatic live utilization, Oregon water quality and metal pollution indices are computed. Based on the obtained results, water of the Qarun Lake is classified as very poor for the fishing utilization, where water quality index changed from 17.05 to 67.4 at the different stations. The metal indices; Pollution index (PI) and metal index (MI); showed that Qarun Lake water undergoes from a several grade of contamination for usage of aquatic life because of the enormous amount of wastes that discharging into the lake. Novel synthesized alginate nanocomposites (alginate, chitosan-alginate, and LDH-alginate and LDH-chitosan-alginate beads) were applied to remove heavy metal such as (Cd^{2+}, Cu^{2+}, Cr^{6+}, Mn^{2+}, Fe^{2+}, Ni^{2+}, Pb^{2+}, and Zn^{2+}) from different wastewater discharging into Qarun Lake. LDH-chitosan-alginate beads the exhibited the maximum removal efficiency that varied between 92.253 and 98.833 % for Cr^{6+} and Fe^{2+}, respectively; in contrast, the alginate beads recorded the lowest one. The study indicated that the nanocomposite adsorbents, especially, LDH-chitosan-alginate beads are an effective substance for metal removal from different wastewater.

Keywords: Alginate nanocomposite beads; Heavy metal removal; Qarun Lake; Pollution index; Metal index; Water quality index.

1. Introduction

Water is an essential substance on which the life depends. Where there is water there is life. Only 0.3% of the freshwater on Earth is consisted of lakes, river systems and reservoirs that included the greatest available water source to satisfy life requirements in our daily lives [1]. On the other hand, Lakes are integrators and mirrors for the effects of climate change and human activity on watersheds, airsheds, and other landscape components. Qarun Lake is a closed ecosystem, receives annually about 400 million cubic meters of drainage water through 12 drains of which 'El-Bats' and 'El-Wadi' are the main drains that transfer most of the wastewater into the lake. The minor drains were recently attached with a huge channel; namely Dayer El-Birka Drain, which transport a lot of wastewater to the lake by pumping. The Egyptian company of salts and Minerals (EMISAL) was built on its coast to extract salts and minerals [2]. The water level in the lake depends on the balance between the inflow of drainage water coming from the cultivated land of Fayoum Province [3] and evaporation rate.

In recent years, the aquatic environment and related issues have been a main concern of the public because most of our ecological water systems are being continuously contaminated. The existence of harmful organic compounds in water supplies and in discharged wastewater is mainly because of the contribution of chemical industries, power plants, landfills, and agricultural sources. Surface runoff may also participate [4]. The result of the poorly water quality affected all living organisms health of an aquatic system [5].

As wastewater discharged without effective treatment into water bodies, hazardous severe effects to environment and ecosystem are increased [6] according to the elevation of heavy metal in waters and sediments causing the destruction of entire aquatic life [7]. Exposing fishes to the high level of trace elements in the polluted aquatic system make and force them to take up metals directly from the environment. This happened through their blood and so with contact with their organs and tissues and accumulated to different degrees [8].

In the past decades, wastewater was treated using traditional treatment methods. There are several treatment processes were available for remediation of industrial wastewater such as biological treatment, coagulation/flocculation, ozone treatment, oxidation, ion exchange, adsorption and photocatalytic degradation [9]. Nanomaterials have facilitated to be applicable in water wastewater treatment [10,11]. The recent advantage of its usage in wastewater remediation usually as it has a higher density of active sites per unit mass. This is because of their larger specific surface area as well as nanomaterials have a great surface free energy. Consequently, it enhanced surface reactivity. By utilizing the size-dependent effects advantage, the current wastewater treatment process could be greatly improved via entering nanomaterials into the system [11]. Abd el Moaty [12] synthesized Co–Fe LDH using the ball milling method that used as an adsorbent for the removal of Cd^{2+} ions from aqueous solutions. That demonstrated 95% Cd^{2+} removal at pH 9 and 6 hrs.which could be applied in wastewater treatment characterized by a high efficiency and low cost. Mahmoud [13] used Novel adsorbent Ti–Fe chitosan LDH intercalated with the nitrate ions that utilized in removing Cd^{2+}, PO_4^{3-}, and benzoquinone from wastewater in single and multiple solutions with agreeable results compared with the previous studies like Moaty [14]. Also, the fabricated LDH showed moderated antimicrobial potency with MIC values ranging from 62.5 to 700 µg/mL as antibacterial agents [13].

The main objective of the current study to evaluate the quality of Qarun Lake water due to the effect of drains wastewater discharging into the lake. In addition, to assess the efficiency of our synthesized nanocompsites for the removal of trace metals from the different wastewaters.

2. Materials and methods

2.1. Study area

Lake Qarun is a remnant of "the historic Lake Moreis" and was originally a fresh water lake. It is a general tank for agricultural wastewater of El Fayoum province. Its surface area is about 243 km^2 with an irregular shape of about 40 Km length and 6.7 Km maximum width, mean depth 4 m and a volume about 1 Billion m^3 and it lies in the arid zone, so it is considered as a tropical lake [15]. The lake is located between longitudes of 30 ° 24` & 30 ° 49` E and latitude of 29 ° 24` & 29 ° 33` N [16]; latitudes at 43-44 m below the sea level. The lake received its water from agriculture waste of El Fayoum province via El-Wadi and El-Bats drains as well as a group of the minor drains that are collected together in Dayer El Birka Drain, the pump station transfer the wastewater from Dayer El Birka Darin into Qarun Lake [17].

2.2. Samples Collection

Six stations were selected to cover the area under investigation beside three main drains; El-Bats, El-Pump station and El-Wadi drains (Figure 1 and Table 1)

Table 1: Details, longitude and latitude of the sampling locations at Qarun Lake.

Code	Name	Latitude (N)	Longitude (E)
Q1	In the far eastern side, in front of El-Bats Drain	29° 30′ 55″	30° 48′ 21″
Q2	In the eastern side, in front of pump station outlet	29° 29′ 35″	30° 46′ 45″
Q3	In the middle sector,	29° 30′ 05″	30° 43′ 05″
Q4	In the middle sector, in front Emisal Company.	29° 28′ 55″	30° 41′ 17″
Q5	In front of El-Wadi Drain.	29° 28′ 15″	30° 39′ 038″
Q6	In the western sector,	29° 26′ 06″	30° 37′ 32″

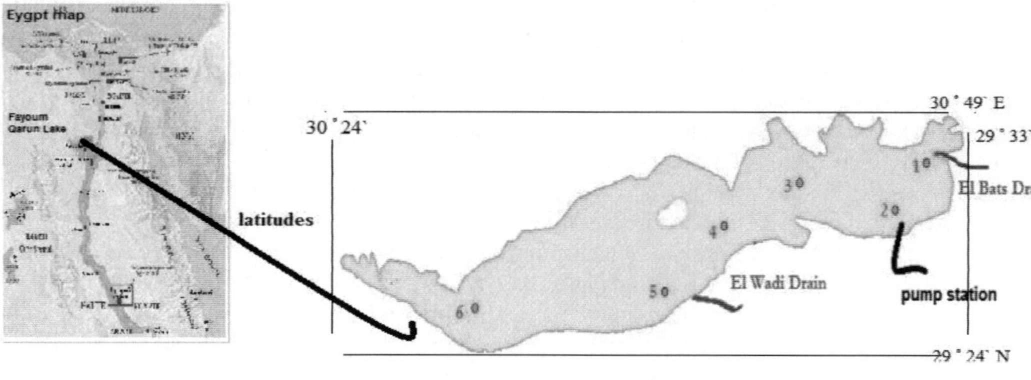

Figure 1. Map showing sample locations along Qarun Lake.

2.3. Methodology of water parameters evaluation

pH, Electrical conductivity (EC), and water temperature were measured *in-situ* using hydro-lab model Orion Research Ion Analyzer 399A, and transparency was estimated via the Secchi disk (diameter 30 cm). Water samples were kept in 2-L polyethylene bottles in an icebox to analyze later in the laboratory. The methods of water analyses were done according to APHA, [16]. Total heavy metals were determined after digestion by conc. HNO_3 using the GBC atomic absorption reader model Savant AA-AAS with GF 5000 graphite furnace.

2.4. Statistics

One-way ANOVA was examined through the Excel-Stat software to analyze the spatial and temporal variations; where $p < 0.05$ and $p < 0.01$ were taken for significance and highly significant levels, respectively. The correlations coefficients (r) between the measured parameters were calculated.

2.5. Oregon water quality index (OWQI).

The quality of Qarun lake water was evaluated according to The Oregon Water Quality Index. The OWQI was computed according to Cude [17] as n in equation (1)

$$\mathbf{WQI} = \left(\frac{\mathbf{n}}{\sum_{\mathbf{i=1}}^{\mathbf{n}} \frac{\mathbf{1}}{\mathbf{S_i^2}}} \right)^{1/2} \quad (1)$$

Where, (WQI) defined as the water quality index result, (n) refer to the number of sub-indices and Si is the sub-index i. the details of calculation of sub-index for each parameter were shown in Cude data [18].

2.6. Metal quality indices

Two different quality indices are used to evaluate the metal contamination of Qarun lake water.

2.6.1. Pollution Index (PI)

It is based on individual metal calculations and categorized to five classes according to equation (2) [19]

$$P_I = \frac{\left([\frac{C_i}{S_i}]^2_{max} + [\frac{C_i}{S_i}]^2_{min}\right)^{1/2}}{2} \quad (2)$$

2.6.2. Metal Index (MI)

It is based on a total trend evaluation of the present status. Elevate concentration of metals as comparing to its respective MAC value, resulted in a worse impact on water quality. As, MI value more than unity is the warning limit [20]. According to Tamasi and Cini [21], the MI is calculated by the equation (3).

$$MI = \sum_{I=1}^{N} \frac{C_i}{(MAC)_i} \quad (3)$$

Where, Ci: is the concentration of each element and MAC: refer to maximum allowable concentration.

2.7. Metal removal with synthesized adsorbents nanocomposites

Four synthesized nanocomposites (alginate, chitosan-alginate, and Ni-AL- LDH-alginate and Ni-Al-LDH-chitosan alginate beads (that were discussed in details in El-Rouby et al., [22]), were assessed to remove the metals from the different wastes that discharge into Qarun Lake.

The treatment process was done for the wastewater by two ways, the first one was achieved to eliminate the metals from the raw wastewater at ambient temperature and normal pH (7.68, 754 and 7.85 for El-Bats, El-Wadi and pump station wastewater, respectively), using 2g/l prepared nanocomposite beads and 300 rpm agitation speed for 2 h.

In the second way, 10 times concentrated samples were treated with the synthesized nanocomposites. For the concentrated samples, mix well 2000 ml in a beaker on a hot plate, add 10 ml of conc. HNO_3 then slow boil, and evaporate, till the lowest volume possible 200 ml and adjust pH using 1N NaOH. The metal removal of the waste was processed at ambient temperature using 2g/l prepared nanocomposite beads, at pH 5 for 2 h contact time and 300 rpm agitation speed.

Generally, The concentration of studied metals (Cu^{2+}, Cr^{6+}, Cd^{2+}, Fe^{2+}, Ni^{2+}, Mn^{2+}, Pb^{2+} and Zn^{2+}) was examined in the different wastewater before and after the removal process using atomic absorption spectroscopy (Savant AA AAS with GF 5000 Graphite Furnace).

3. Results and discussion

The standard deviations, mean values, and ranges of the obtained results are presented in tables 2&3.

3.1. Physical characteristics

Temperature is an important factor in the aquatic environment since it affects directly or indirectly not only on the survival and distribution of the aquatic species at different stage of life, but also on their development, growth rate, activity, reproduction and make it susceptibility to many diseases [23]. It ranged between the lowest value 15.40 ºC during winter and the highest one 30.50 ºC during summer with highly temporal significant difference ($p < 0.01$). The variation in water temperature depends mainly on the climatic conditions, sampling times, the number of sunshine hours and affected by specific characteristics of water environment such as turbidity, wind force, plant cover and humidity [24]. Temperature is positively correlated (n=24, P<0.05) with BOD (r=0.51), While it is negative correlated (n=24, P< 0.01) with COD (r=-0.85).

Water transparency evaluates the depth of the photic zone and thus affects the lower limit of light penetration that influences the primary productivity of a lake. Plankton and suspended particles reduce transparency in natural waters. Transparency was fluctuated between 20 and 100 cm with spatial significant difference (p< 0.01). Generally, Qarun Lake is characterized by water turbidity due to the deteriorating effect of El-Bats and El-Wadi drains [25]. The sites in front of the drains exhibited the lowest transparency values. On the other hand, the western part of the lake exhibited the highest values of transparency as confirmed by Fishar et al [26]. Transparency is a high negatively correlated (n= 24, p< 0.01) with TSS (r = -0.61), Fe^{2+} (r = -0.66), Mn^{2+} (r = -0.77), Ni^{2+} (r = -0.60), Cr^{6+}(r=-0.59), NH_4 (r=-0.81), NO_2 (r=-0.73) and NO_3(r=-0.72), TP(r=-0.79) and PO_4 (r=-0.72).

Water electrical conductivity (EC) is the ability of water to conduct an electrical current, and the dissolved ions are the conductors [27].Where the higher value of conductivity, the greater ions and higher dissolved solids. In Qarun Lake water, EC was fluctuated between 17.09 and 47.24 ms/cm with spatial significant difference (p< 0.01). The stations that are facing the drains (El-Bats, pump station and El-Wadi drains) exhibited the lowest E.C values than other lake station. This may be related to the direct dilution effect of the drainage water. E.C is a high positively correlated (n=24, p<0.01) with TS(r=0.98), Salinity (r=0.98) as agreed with many authors [24]. While, it is negatively correlated with high significance (n=24, p<0.01) with TSS (r=-0.66) and NH_4 (r=-0.81).

Total solids denote to any substance either suspended (TSS) or dissolved (TDS) in water. It described the chemical contents of Lake water to evaluate edaphic relationships of aquatic environment that affected productivity within the aquatic system [23]. TS, TDS, and TSS in Qarun Lake varied in the range of (13.19-37.82)g/l, (13.22-37.84)g/l and (13.60-39.26) mg/l, respectively. TDS and TS show a spatial significant difference (p< 0.01). The maximum salinity (TDS) and TS value have observed during summer due to temperature elevation and evaporation rate. TS and TDS are positively correlated (n= 24, p< 0.01) with pH (r=0.81) in addition to the most major ions, that is in agreement with Goher [24]. The corresponding values in the drains water fluctuated in the ranges (2.03-3.78) and (1.99-3.74) g/l, which are much lower than the lake water, while TSS showed a remarkable increase in the drains wastewater and ranged between (31.98-52.16) mg/l.

3.2. Chemical parameters

The pH measures the acidity or alkalinity of water. A level of pH is an important factor in the chemical and biological systems of natural waters [28]**.** pH of Qarun Lake water lies in the alkaline side and it varied in the ranges of (7.97-8.95). The maximum pH~8.97 was recorded at station (3) during summer, while the minimum pH~7.67 was recorded at station (5) during autumn. ANOVA study show a high spatial significant difference (p< 0.01). However, in the drainages water pH ranged between 7.16 and 8.02. pH is positively correlated (n= 24, p< 0.01) with E.C (r=0.8), anions like SO_4^{2-} (r=0.79) and cations like Na^+ (r=0.80) and Mg^{2+} (r=0.79). While it is negatively correlated with heavy metals as Mn (r=-0.67), Ni (r=-0.73) and Cr (r=-0.69).
Dissolved oxygen (DO), Chemical Oxygen Demand (COD) and Biological Oxygen Demand (BOD) values varied in the ranges of (6.20-10.62), (18.78-30.12) and (4.10-8.66) mg/l. The lowest DO value was observed at station 5 in front El-Wadi drain, which may be related to the impact of the huge amount wastewater loaded with the organic matter. Moreover, the maximum COD and BOD values were recorded in summer, which may be attributed to the higher rate of decomposition of organic substance under the effect of the high temperature and bacterial activity, this finding is in a harmony with those obtained by both Abdel-Satar et al [2] and Goher [24]. COD showed a temporal significant difference (p< 0.01). Nevertheless, DO and BOD values exhibited a spatial significant difference (p< 0.05). The coefficient correlation data showed that COD is negatively correlated with temperature (r=-0.85, P <0.01), however, BOD is positively correlated (n=24, P<0.5) with temperature (r=0.51) and DO (r=0.48). DO is negatively correlated to most nutrient as ammonia NH4 (r=-0.45, P<0.05) and

with metals as Cr (r=-0.45), Mn (r=-0.42) and Ni (r=-0.66, $p<0.01$) which indicate the role of high dissolved oxygen in the settling of metal into the sediment, these results are in agreement with Abdel-Satar et al [2]. On the other hand, DO, BOD and COD of drains water were in the ranges of (2.37-8.12) mg/l, (4.43-7.67) mg/l and (15.46-21.80) mg/l, respectively.

Table 2. Range and mean of Qarun Lake water parameters compared to guidelines of aquatic life protection.

parameter	Range	Mean±SD	Aquatic live[a]
Temp. (°C)	15.3-31.7	21.58±5.47	8–28
Transparency (cm)	20-100	57.71±20.16	
EC (ms/cm	17.09-47.24	36.37±8.43	
TDS (g/l)	13.19-37.82	29.59±7.262 [b]	
TSS (mg/l)	13.60-39.26	24.32±6.34	25
TS (g/l)	13.22-37.84	29.62±7.25 [b]	
pH	7.97-8.95	8.40±0.26	6.5–9
DO (mg/l)	6.20-10.62	8.32±1.10	>5.5
BOD (mg/l)	4.10-8.66	6.55±2.02	< 6
COD (mg/l)	18.78-30.12	24.10±3.29	
CO3(mg/l)	9.12-16.21	12.66±2.39	
HCO3(mg/l)	292.0-347.0	317.24±14.89	
Cl (mg/l)	4.65-13.31	10.73±2.54[b]	
SO4(mg/l)	3.42-9.88	7.87±1.88 [b]	
Ca (mg/l)	0.16-0.52	1.17±1.49 [b]	
Mg (mg/l)	0.47-1.35	1.07±0.26 [b]	
Na (mg/l)	3.78-10.93	8.68±2.11 [b]	
K (mg/l)	0.08-0.24	0.19±0.047 [b]	
NO3-N (mg/l)	0.03-1.06	0.30±0.36	2.93
NO2-N (lg/l)	0.00-288.90	60.55±84.12	60
NH4-N (mg/l)	0.05-2.15	0.55±0.59	1.37
TN	0.86-5.14	2.49±2.54	
PO4-P (lg/l)	1.10-78.42	26.08±22.36	
TP (lg/l)	32.77-232.0	91.97±59.77	
SiO3(mg/l)	4.48-0.44	6.15±0.74	
Fe (μg/l)	111.91-481.24	205.34±95.81	300
Mn (μg/l)	11.00-40.85	20.51±6.93	50
Zn (μg/l)	13.60-38.36	27.30±6.39	50
Cu (μg/l)	4.09-10.32	7.00±1.78	4
Ni(μg/l)	3.59-12.65	7.90±2.70	25
Cr (μg/l)	7.06-17.06	10.43±2.62	10
Cd(μg/l)	0.43-1.49	0.99±0.28	1
Pb (μg/l)	11.87-36.41	20.28±5.56	7

[a] Canadian Council of Ministers of the Environment (CCME) [29].
[b] The average of anions and cations of the lake water is higher if the stations facing the drains are excluded

3.2.1. Major Ions

Although the salinity value of Qarun Lake is similar to the salinity of sea water. But the ionic composition of Qarun Lake water differ completely from the composition of sea water, due to the different source of salts which discharging to the oceans and Qarun Lake [24]. This observation is based mainly on the chloride and sulphate levels. Sulphate in Qarun Lake water up to about 10 g/l, which is considerable higher than in sea water 2.65 g/l. on the other side, chloride in Qarun Lake water is lower than that in sea water, where it recorded about 14 and 19.35 g/l, in Qarun Lake and sea respectively.

In the current study, carbonate, bicarbonate, chloride and sulphate values were in the ranges of (9.12-16.21) mg/l, (292.0-347.0) mg/l, (4.65-13.31) g/l and (3.42-9.88) g/l. with a highly spatial significant difference ($p< 0.01$). Generally, the increase of the anions concentrations in summer may be related to the evaporation effect, Also, the increase of carbonate levels during summer may be attributed to photosynthetic activity in addition to the elevation of water temperature that leads to decrease in solubility of carbon dioxide [30]. On the other hand, carbonate in the drains wastewater was completely depleted, while bicarbonate, chloride, and sulphate were varied in the ranges of (352.0-422.0) mg/l, (0.46-1.20) g/l and (0.28-0.59) g/l, respectively.

Table. 3. Range and mean of wastewater parameters discharching into Qarun Lake during (2015).

drain	El -Wadi drain		El- Bats Drain		pump station	
parameter	Range	Mean±SD	Range	Mean±SD	Range	Mean±SD
Temp. (°C)	15.40-30.50	21.68±6.59	15.60-29.40	21.23±6.04	15.60- 30.50	21.73±6.52
Transparency (cm)	5.0-20.0	15.00±7.07	10.0-30.0	22.50±8.66	25.0-35.0	31.25±4.79
EC (ms/cm	3.17-4.91	3.99±0.73	2.78-5.28	3.76±1.11	2.54-3.33	3.01±0.33
TDS (g/l)	2.26-3.40	2.79±0.47	1.99-3.74	2.61±0.79	2.25-3.39	2.78±0.47
TSS (mg/l)	41.86-52.16	44.66±5.00	38.34-42.00	40.22±1.54	31.98-37.67	35.04±2.83
TS (g/l)	2.30-3.44	2.83±0.47	2.03-3.78	2.65±0.79	2.28-3.43	2.81±0.48
pH	7.16-7.91	7.64±0.35	7.38-7.97	7.66±0.25	7.67-8.02	7.87±0.145
DO (mg/l)	2.37-7.81	5.26±2.38	2.59-6.67	4.58±1.74	6.08-8.12	7.02±0.84
BOD (mg/l)	4.43-7.67	6.16±1.38	4.85-6.82	5.66±0.88	5.42-6.62	5.97±0.64
COD (mg/l)	15.88-19.16	18.18±1.55	17.62-21.80	19.32±1.93	15.46-18.24	16.38±1.26
CO3(mg/l)	ND		ND		ND	
HCO3(mg/l)	371.0-422.0	393.25±22.68	396-411	390.75±75	352.0-392.0	375.75±16.94
Ca (mg/l)	0.04-0.08	0.05±0.014	0.04-0.09	0.06±0.021	0.03-0.05	0.04±0.008
Mg (mg/l)	0.04-0.07	0.05±0.013	0.04-0.08	0.06±0.02	0.04-0.05	0.04±0.006
Na (mg/l)	0.58-0.88	0.74±0.12	0.50-0.97	0.68±0.21	0.46-0.59	0.54±0.05
K	0.02-0.04	0.22±0.008	0.02-0.04	0.03±0.006	0.02-0.03	0.02±0.005
NO3-N (mg/l)	0.75-1.49	0.04±0.01	0.81-1.69	1.4±0.41	0.22-0.84	0.52±0.27
NO2-N (µg/l)	244.36-618.2	456.94±182.94	174.60-466.46	332.55±120.34	158.5-222.12	182.86±27.76
NH4-N (mg/l)	2.47-3.39	2.98±0.39	0.81-2.57	1.94±0.79	0.50-1.59	1.20±0.49
TN	3.94-6.70	5.18±1.4	2.74-6.26	4.99±1.6	2.11-5.29	3.68±1.3
PO4-P (µg/l)	122.1-259.6	172.7±61.32	107.8-195.8	162.53±38.45	76.24-92.15	83.22±8.17
TP (µg/l)	263.86-489.6	339.91±105.389	222.76-406.11	331.77±77.50	164.22-192.64	175.67±12.45
SiO3(mg/l)	6.98-9.84	8.21±1.2	7.18-10.42	9.18±1.4	5.98-8.62	9.18±1.4
Fe (µg/l)	358.95-613.6	490.78±124.3	378.307±592	513.39±98.3	307.35-586.1	416.18±119.4
Mn (µg/l)	27.52-77.50	46.50±21.58	42.86-62.10	51.73±9.19	21.79-77.49	42.70±25.65
Zn (µg/l)	31.92-55.78	42.31±10.02	36.58-50.82	42.04±6.65	38.31-60.17	47.31±9.26
Cu (µg/l)	9.00-18.38	13.48±4.53	11.29-19.08	15.21±3.80	9.58-15.38	12.50±2.99
Ni(µg/l)	17.29-20.80	18.18±1.48	13.97-18.35	16.34±1.98	12.33-23.25	18.36±4.57
Cr (µg/l)	22.15-49.84	35.06±13.34	14.81-34.56	23.87±8.16	16.10-35.75	25.30±9.74
Cd (µg/l)	0.82-2.47	1.82±0.71	0.86-2.26	1.56±0.61	0.53-2.10	1.60±0.725
Pb (µg/l)	23.32-53.00	31.67±14	29.74-44.96	38.75±6.43	20.52-48.92	34.95±11.76

The major cations; calcium, magnesium, sodium and potassium varied in the ranges of (0.16-0.52) mg/l, (0.47-1.35) mg/l, (3.78-10.93) g/l and (00.08-0.24) mg/l, respectively, with highly spatial significant difference ($p < 0.01$). The results revealed that Ca^{2+}, Mg^{2+}, Na^{+} and K^{+} have the same distribution pattern of Cl- and SO_4^{2-}, where it increased westwards and decreased at the stations close to the discharging points of the drains. The corresponding values in the drains fluctuated between (0.03-0.09) mg/l, (004-0.08) mg/l, (0.46-0.58) g/l and (0.02-0.04) mg/l for Ca^{2+}, Mg^{2+}, Na^{+} and K^{+}, respectively.

Carbonates is negatively correlated (n= 24, $p < 0.05$) with heavy metals as Fe^{2+} (r=-0.48), Cu^{2+} (r=-0.48) and Pb^{2+} (r=-0.42), which indicated the precipitation of these metals as carbonate. Bicarbonates is positively correlated (n= 24, $p < 0.01$) with Ni^{2+} (r=0.55), (n=24, $p<0.05$) with Zn^{2+} (r=0.44), TN (r=0.44), TSS(r=0.48), and temperature (r=0.47). The high positive correlation (n= 24, $p < 0.01$) between chloride and sulphate with the major cations (r=0.99) indicated that a large amount of the major cations was presented in chloride and sulphate forms, these findings are in a consensus with the results those obtained by Goher et al [31].

3.2.2. Nutrient Salts

Nutrients of concern with respect to water quality are nitrogen and phosphorous. Excess nitrogen and phosphorous in lakes and streams can cause algal blooms and eutrophication [32]. Nutrient in surface water is connected to the natural conditions and anthropogenic effect [31]. Ammonia, nitrites, nitrate and total nitrogen were found in the ranges of (0.05-2.15) mg/l, (0.00-288.90) µg/l, (0.03-1.06) mg/l and (0.86-5.14) mg/l, respectively. Ammonia, Nitrite and Nitrate show a high significant difference ($p<0.01$) among the seasons and sites. While TN values show a temporal significant difference ($p< 0.05$).

Phosphorous is a nutrient limited and needed for the growth of all plants, aquatic plants and Phytoplankton alike. Once phosphorus accumulates within a lake, it can cycle through the water column and promote algal blooms indefinitely [33]. Phosphorus forms showed an irregular distribution pattern, with a high spatial significant difference ($p<0.01$). Orthophosphate and TP varied in the ranges of (1.10-78.42) and (32.77-232.0) µg/l, respectively. These results declared that Qarun Lake is considered a eutrophic system; Mueller and Helsel [34] cited that a lake with a concentration exceeding 0.020 mg of phosphorus are already considered eutrophic.

Phosphorous and nitrogen are important for the conservation of the productivity of the lake; conversely, the high levels cause algal blooms, harmful to most aquatic organisms [35], which leads to oxygen depletion and cause fish mortality [31].

Reactive Silicate fluctuated between 2.11-6.70 mg/l with a spatial and temporal significant difference ($p<0.01$ and $p<0.05$, respectively). According to Abdel-Satar et al [36], the dissolution of sediments and silicate rocks in addition to the absorption by diatoms control the silica levels. Generally, the high concentrations of the nutrient salts were recorded in the stations in front the discharging point of the drains, especially stations 5 and 1.

The corresponding values in the drains were found much higher than the lake and changed in the ranges of (1.20-0.49) mg/l, (0.50-3.39) µg/l and (0.22-1.69) mg/l, (2.11-6.70) mg/l, (76.24-259.6) µg/l, (164.22-489.67) µg/l and (5.98-10.42) mg/l, for ammonia, nitrites, nitrate, total nitrogen, orthophosphate, total phosphorus and reactive silicate, respectively.

The obtained results showed a high positive correlation among the different nutrient forms ($r = 0.68- 0.93$ $n=24$, $p<0.01$). Moreover, the positive correlation with different heavy metals ($r = 0.59 - 0.73$ $n=24$, $p<0.01$). These observations are in compatibility with that reported by both Goher et al and Abdel-Satar et al [31,36] and indicate that the different wastes are the common source of these parameters.

3.2.3. Heavy metals

Heavy metals not only enter the aquatic ecosystem through erosion, atmospheric deposition and anthropogenic activities but also it increased due to industrial effluents, domestic sewage, and mining [37]. So, presence excess of them in the water environment, especially lakes and rivers caused a hazard to humans by their toxicity and potential pollution to the food chain. Unlike many organic wastes, trace metals cannot be converted biologically into harmless products [38].

The results of the studied metals (iron, manganese, zinc, chromium, copper, lead, nickel and cadmium) showed high significant variations among the locations and seasons. Where the highest contents were recorded at the stations facing the outlet of the drains. Otherwise, the metal levels in Qarun Lake water have increased during the hot period, which may be accredited to the metal liberation from the bottom sediment to the water column due to the degradation of organic matter under the effect of the high temperature and the microbial activity [9]. Inversely, the lower metals content in water during cold period; which is witnessing algae blooms; may be due to uptake of it by phytoplankton and the dead phytoplankton settling to the bottom sediment leaving the water poor in the trace elements, in addition to settling of metals from water column to the sediments under slightly high pH values [24]. The results declared that the levels of Fe^{2+}, Cr^{6+} and Cd^{2+} in Qarun Lake water exceeded the guidelines for aquatic-life of CCME [29].

Iron and manganese are the greatest common metals found in nature. They varied in Qarun Lake water between 111.91-481.24 and 11.00-40.85 μg/l, respectively. Iron and manganese concentrations in surface water are mainly controlled by redox conditions, the type and amount of dissolved organic matter and water pH. The bio-essential element iron is not toxic itself, but it enhances the uptake of As (III) in some aquatic organisms and inhibits those of As (V) [39]. Further, zero valiant iron has been tested by Gheju, [40] as a promoter of the reduction of the heavily toxic and mobile Cr (VI) into the less toxic Cr (III), a process which depends mainly of the conditions of the environmental of the aquatic system.

According to Ruttner [41], the distribution dynamics of iron in the aquatic system depended on DO and is usually present as insoluble $Fe(OH)_3$ or ferric oxide. The ferrous form can only exist in the depletion of oxygen and the ferric form is almost completely insoluble. In other words, at oxygenated water iron is precipitated as ferric form, while in the insufficiency of oxygen and the water containing CO_2 comes in contact with iron in the ferrous form, in this case, ferrous bicarbonate goes into solution; however, at the oxidation occur insoluble ferric hydroxide ions precipitated as follows:

$$4Fe(HCO_3)_2+2H_2O+O_2 \overset{Yield}{\Longleftrightarrow} 4Fe(OH)_3+8CO_2 \quad (4)$$

Also, in water contains ferrous sulfate which is unstable in presence of dissolved oxygen, ferrous sulfate and sulphuric acid (formed by the action of microorganisms) react under the presence of oxygen to produce ferric sulfate, which is quickly hydrolyzed to ferric hydroxide. Manganese is very close to iron in the economy of rivers, lakes and behave in much the same manner. The chemistry of manganese is appreciably complicated due to its existence in different oxidation states [24].

The distribution dynamics of zinc, chromium, lead, copper nickel and cadmium follow the distribution of manganese and iron and were varied in ranges of (13.60-38.36), (44.09-10.32), (3.59-12.65), (7.06-17.06), (11.87-36.41) and (0.43-1.49) μg/l, respectively. The positive significant correlations ($r= 0.42$–0.71; $n =24$, $p< 0.01$) among each pairs of the heavy metals indicated the association behavior and the common sources. Also, the statistical analysis showed a negative correlation ($n=24$, $p<0.01$) between the most studied metals with dissolved oxygen ($r = -0.42$: -0.66) and with pH ($r = -0.78$: -0.75), which indicate the effect of high dissolved oxygen and pH in the precipitation of the different metals into the bottom sediment these findings are in agreement with many studies [2,24].

3.3. Water quality index (WQI)

The quality of Qarun Lake water was investigated according to Oregon Water Quality Index (OWQI) module, which was developed by the Oregon Department of Environmental Quality (ODEQ), for the purpose of summarizing and evaluating water quality trends and status, in the late 1970s and updated several times since then [42].The Oregon Water Quality Index is a single number that express water quality by integrating values of 8 parameters (temperature, dissolved oxygen, BOD, pH, (ammonia + nitrate nitrogen), total phosphorus, total solids and fecal coliform). The OWQI comforts to estimate the efficiency of the management activities of the water quality. It may also be to progress ecological indicators, such as the percentage of river and lakes monitoring sites with significantly improving water quality, or the percentage of sites with excellent water quality [42]. OWQI is classified to 5 categories (Table 4) [17].

In the current study, six parameters were used to calculate the OWQI for fishing usage. The variables included temperature, DO, BOD, pH, (ammonia $^+$ nitrates) and total phosphorus. The obtained results indicate that WQI for Qarun Lake water changed from 17.05 to 67.4 at the different stations. Therefore, Qarun Lake water during all seasons is classified as very poor for the Fishing utilization at all stations, except station 6 that is categorized as poor in winter and spring (Figure 2).

3.4. Metal quality indices

Two different indices are designed to evaluate the metal contamination of Qarun Lake water, the first one is the Pollution Index (PI), that independent on the effect of the individual metal. While the other index, Metal index (MI), based on the effect of the total metals level. In the present study, the acceptable limits of the trace metals obtained by CCME [29] for aquatic life protection were used as maximum allowable concentration (MAC) value.

Table 4. Categories of OWQI

OQWI Value	Class
10-59	Very poor
60-79	Poor
80-84	Fair
85-89	Good
90-100	Excellent

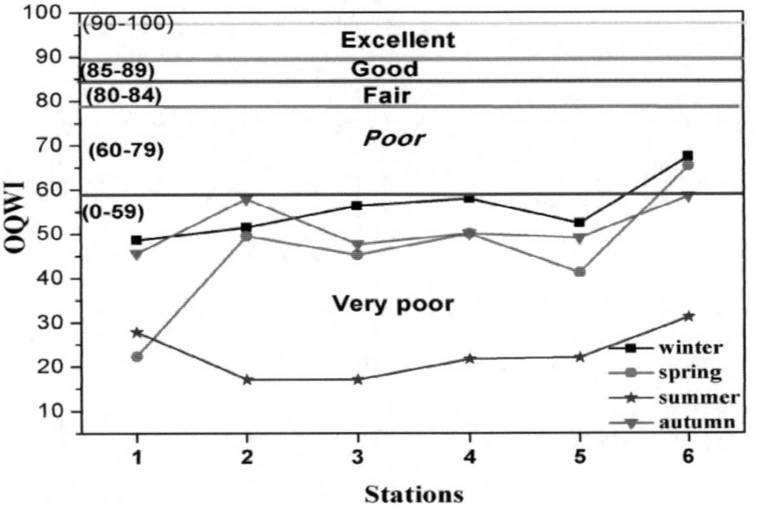

Figure 2. OWQI and its categorizations of Qarun Lake water for fishing utilizations.

3.4. 1. Pollution Index (PI)

The pollution index (PI) was used in our study to determine the grade of trace metals toxicity in water samples. It is based on individual metal calculations and categorized to five classes (Table 5) according to Caeiro et al [43].

The results indicate a various pollution degree of the studied metals for aquatic life in Qarun Lake. Pollution index exhibited that Mn, Ni and Zn have not polluted effect for aquatic life usage. On the other hand, Cu and Cr showed slight pollution to moderate effect at all station. Cd showed slightly pollution effect at stations 1, 5 and 6, while Fe showed the same effect at stations 1 and 4. Finally, Pb cause strongly pollution effect at all stations (Table 6).

Table 5. Categories of water pollution index

Class	PI Value	Class
1	≤ 1	No effect
2	>1-2	Slightly affected
3	>2-3	Moderately affected
4	>3-5	Strongly affected
5	>5	Seriously affected

3.4.2. Metal Index (MI)

It is based on a total trend evaluation of the present status. By increasing concentration of a metal compared to its respective maximum allowable value (MAC), the worse impact of the quality of the water appeared. MI value >1 is the warning limit [18]. According to MI values, all selected sites in Qarun Lake are suffer from seriously metal pollution for the Aquatic life utilization (Table 7).

Table 6. Pollution index of the measured metals in the Qarun Lake water according to guideline levels of aquatic life utilizations.

Station	Fe		Mn		Zn		Cu	
	PI value	effect	PI value	Effect	PI	Effect	PI value	Effect
1	1.14	slightly	0.67	No	0.83	No	2.14	moderatel
2	0.56	No	0.43	No	0.71	No	1.96	slightly
3	0.69	No	0.31	No	0.57	No	1.53	slightly
4	0.67	slightly	0.38	No	0.55	No	1.56	slightly
5	1.18	No	0.70	No	0.55	No	2.18	moderatel
6	0.49	No	0.31	No	0.48	No	1.77	slightly
Station	Ni		Cr		Pb		Cd	
	PI value	effect	PI value	Effect	PI	Effect	PI value	Effect
1	0.53	No	2.18	moderatel	3.47	strongly	1.27	slightly
2	0.41	No	1.74	slightly	3.54	strongly	0.97	No
3	0.30	No	1.84	slightly	3.39	strongly	0.99	No
4	0.27	No	2.08	moderatel	3.99	strongly	0.90	No
5	0.40	No	2.11	moderatel	4.03	strongly	1.13	slightly
6	0.42	No	1.67	slightly	3.90	strongly	1.09	Slightly

Table 7. Metal index of the measured metals in Qarun Lakes water for aquatic life utilizations.

Stations	MI value	Rank
1	11.73	Polluted
2	9.78	Polluted
3	9.07	Polluted
4	9.88	Polluted
5	10.96	Polluted
6	8.44	Polluted

3.5. Application of removal of toxic metal from wastewater discharged into Qarun Lake using synthesized nanocomposite

In the present study, the prepared Ni-Al-CO_3 layered double hydroxide (LDH) with Ni:Al ratio(3:1) based nanocomposites with alginate and chitosan beads were assessed to remove the metals from the different wastes that discharge into Qarun Lake. The characterization of the synthesized nanocomposites, that previously utilized for removal trace metal from aqueous solution and the optimum conditions of these materials for the its adsorption process have discussed in details in El-Rouby et al., [22].

For the raw wastewater, the results revealed that all nanocomposites (alginate, chitosan alginate, LDH alginate and LDH-chitosan- alginate beads) recorded 100% removal efficiency for all studied metals; (Fe^{2+}, Mn^{2+}, Pb^{2+}, Zn^{2+}, Cu^{2+}, Cd^{2+}, Cr^{6+} and Ni^{2+}) for El-Bats Drain, El-Wadi Drain and pump station (Table 8).

On the other hand, the application of heavy metal removal using nanocomposites from the concentrated sample of raw wastewater declared that LDH-chitosan-alginate beads recorded the highest removal efficiency for all studied metals from the different wastes. Where, iron exhibited the most adsorbed ion with a removal ratio of 98.587, 98.765 and 98.333 % for El-Bats, El-Wadi and pump station wastewater respectively by LDH-chitosan-alginate beads, while nickel recorded the

lowest one with 82.588, 85.819 and 85.70 % removal ratio using alginate beads from for El-Bats, El-Wadi and pump station wastewater, respectively (Figure 3).
According to the total ions that were adsorbed with the different adsorbents, the adsorption capacity of the prepared nanocomposites were in the order of LDH-chitosan-alginate > chitosan-alginate > LDH- -alginate >alginate beads. LDH-chitosan-alginate beads exhibited the maximum adsorption capacity (qe) 5.056 mg/g for El Bats Drain wastewater; in contrast, the alginate beads recorded the lowest one of 2.987 mg/g for the pump station wastewater (Table 9). Our studies are in agreement with Rahmanian et al., [44], and Asiabi et al., [45] that confirmed that LDHs proved high removal efficiency for heavy metals.
It is worth mentioning that the synthesized nanocomposites considered an excellent adsorbent to remove different heavy metals from the various wastewaters. All prepared nanocomposites (alginate, LDH-alginate, Chitosan alginate and LDH-Chitosan-alginate beads) give efficiency removal reached to 100% of raw wastewater in all studied drains as illustrated in table.8. However, by application of these adsorbents on the concentrated wastewater samples, the removal efficiency was in the range of (98.833-92.353), (97.835-90.235), (97.039-88.981) and (94.456-82.588) % for LDH-chitosan-alginate, chitosan-alginate, LDH-alginate and alginate beads, respectively (Table 9).
The total adsorption capacity (q_e) mg/g of the prepared adsorbents for all heavy metals removed by these adsorbents was found in the ranges of (4.568-2.987), (4.773-3.143), (4.858-3.210) and (5.056-3.319) for alginate, LDH-alginate, chitosan-alginate and LDH-chitosan-alginate, respectively in all wastewater samples as represented in table.10.

Table 8. The Removal efficiency % of the metals by the synthesized nanocomposites from the raw wastewater.

metal / Drain	Adsorbent	Fe	Pb	Cu	Zn	Cd	Mn	Ni	Cr
El Bats Drain Water	Concentration μg/l	657	57	21.8	56.6	1.8	65.2	44.8	45.3
	Alginate	100	100	100	100	100	100	100	100
	Chitosan- Alginate	100	100	100	100	100	100	100	100
	LDH/ Alginate	100	100	100	100	100	100	100	100
	LDH/chitosan/ Alginate	100	100	100	100	100	100	100	100
El Wadi Drain Water	Concentration μg/l	519	53.94	18.63	52.11	1.63	50.36	37.53	31.06
	Alginate	100	100	100	100	100	100	100	100
	Chitosan- Alginate	100	100	100	100	100	100	100	100
	LDH/ Alginate	100	100	100	100	100	100	100	100
	LDH/chitosan/ Alginate	100	100	100	100	100	100	100	100
pump station	Concentration μg/l	429	46.76	16.77	40.91	1.63	54.32	20.41	25.64
	Alginate	100	100	100	100	100	100	100	100
	Chitosan- Alginate	100	100	100	100	100	100	100	100
	LDH/ Alginate	100	100	100	100	100	100	100	100
	LDH/chitosan/ Alginate	100	100	100	100	100	100	100	100

Figure 3 illustrates the removal efficiency of heavy metals of concentrated samples (ten times) from (a) El-Bats Drain, (b) El-Wadi Drain and, (c) Pump station that discharged into Qarun Lake using different alginate nanocomposites beads. The results demonstrate that the removal efficiency of the prepared adsorbents are in the order of LDH-chitosan-alginate > chitosan-alginate > LDH-alginate > alginate beads. On the other hand, the removal of iron by LDH-chitosan-alginate beads recorded the highest one up to 98.587, 98.765 and 98.833 for the concentrated wastewater of El Bats, El Wadi and pump station drains, respectively. While, the lowest removal efficiency values obtained for Ni^{2+} using Alginate were 82.588% for El-Bats < 85.752 % for pump station drain < 85.819 % for El-Wad Drain.

In the present study, the prepared adsorbents were assessed to eliminate the trace metal from the raw wastewaters and 10 times concentrated wastewater samples of the different drains discharging into Qarun Lake under the finest condition of ambient room temperature, 2g/l adsorbent dose, 2h contact at 300 rpm agitation speed.

Table 9. The Removal efficiency % of the metals by the synthesized nanocomposites from the concentrated wastewater.

Drain	Adsorbent / metal	Fe	Pb	Cu	Zn	Cd	Mn	Ni	Cr
El Bats Drain Water	Concentration μg/l	7220.0	620.00	226.00	621.80	18.90	690.00	473.00	482.70
	Alginate	88.698	86.048	89.553	88.437	87.778	90.116	82.588	86.418
	Chitosan- Alginate	94.155	94.984	92.965	93.036	93.704	95.158	90.235	90.870
	LDH/ Alginate	92.641	92.339	90.664	90.962	92.434	93.420	88.981	89.584
	LDH/chitosan/ Alginate	98.587	96.952	96.062	95.529	95.238	97.212	94.497	92.353
El Wadi Drain Water	Concentration μg/l	5500.0	580.00	194.00	553.70	17.20	530.00	393.00	326.00
	Alginate	88.291	91.845	92.629	90.946	88.721	91.485	85.819	87.847
	Chitosan- Alginate	94.891	95.793	97.835	95.720	94.128	95.019	90.659	92.595
	LDH/ Alginate	92.782	93.931	96.959	94.618	92.791	92.906	89.870	91.525
	LDH/chitosan/ Alginate	98.765	98.552	98.608	97.773	95.756	97.079	94.631	94.433
pump station	Concentration μg/l	4560.0	490.00	179.00	431.10	17.00	570.00	218.00	271.10
	Alginate	87.105	92.469	93.799	91.048	88.824	94.456	85.752	91.479
	Chitosan- Alginate	95.022	96.449	97.709	96.381	94.353	96.298	92.922	94.419
	LDH/ Alginate	92.434	96.082	97.039	95.662	94.059	95.719	91.225	93.298
	LDH/chitosan/ Alginate	98.833	98.776	98.659	97.926	96.000	97.349	95.133	95.876

Table 10. Adsorption capacity (q_e) mg/g of the prepared adsorbents for the different wastewater.

Adsorbent	El Bats Drain	El Wadi Drain	Pump station
Alginate	4.568	3.598	2.987
Chitosan alginate	4.8578	3.836	3.210
LDH alginate	4.773	3.760	3.143
LDH chitosan alginate	5.056	3.976	3.319

The obtained experimental results were confirmed with FTIR spectra for the LDH-chitosan-alginate bead before and after the adsorption process. Figure 4. (a-c) and table 11 displayed FTIR spectra of LDH-chitosan-alginate bead before and after adsorption of Cu^{2+} and Cd^{2+} cations. Where, the shifts and changes in the intensity of characteristic bands of the LDH-chitosan bead spectrum after adsorption -as compared with the spectra of it before adsorption- can be counted on with regard to confirming the interaction processes during adsorption, which is indicative of deprotonation of the functions group. The band at 3418 cm^{-1} declined to 3421, 3428.6 cm^{-1} after adsorption of Cu^{2+} and Cd^{2+}, respectively, suggesting that the bonded OH groups exerted a significant role in Cu^{2+} and Cd^{2+} adsorptions. Appearance of some new peaks at 2379 cm^{-1} and 2171 cm^{-1} for cupper and cadmium, respectively as confirmed with Mahmoud et al [13] that referred to presence of υ(−C=O) of the amide group CONHR of the chitosan and aliphatic groups, specifically CH_2 and CH in cadmium and metal adsorption.

IOP Conf. Series: Materials Science and Engineering **464** (2018) 012003 doi:10.1088/1757-899X/464/1/012003

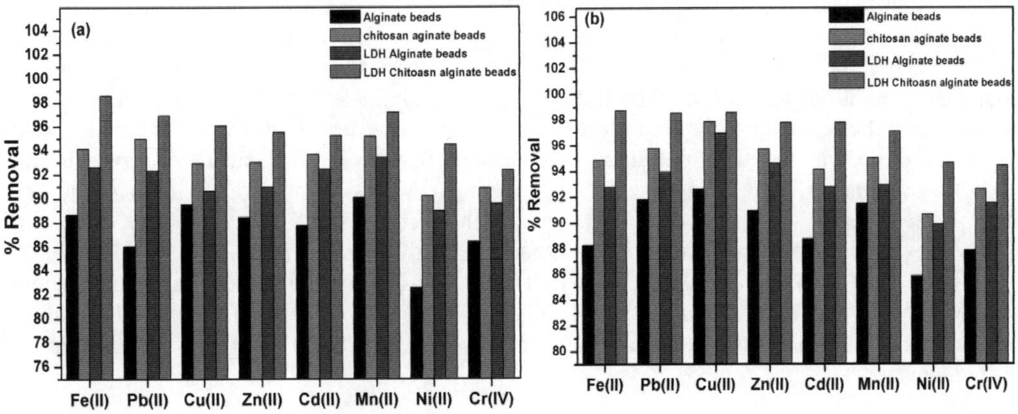

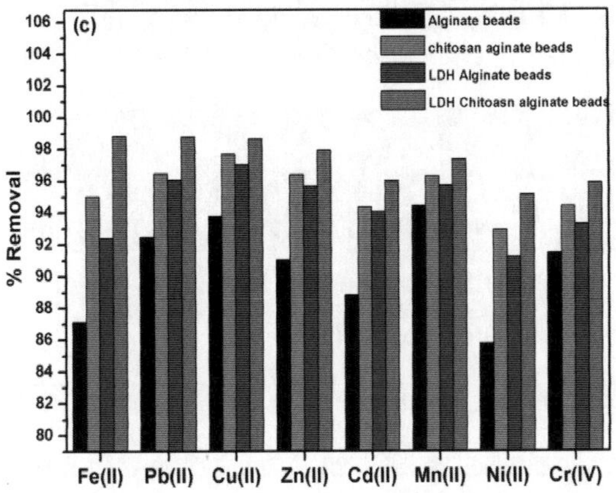

Figure 3. Efficiency removal of heavy metals of concentrated samples (ten times) of (a) El-Bats Drain, (b) El-Wadi Drain and, (c) Pump station discharged on Qarun Lake using different alginate nanocomposites beads.

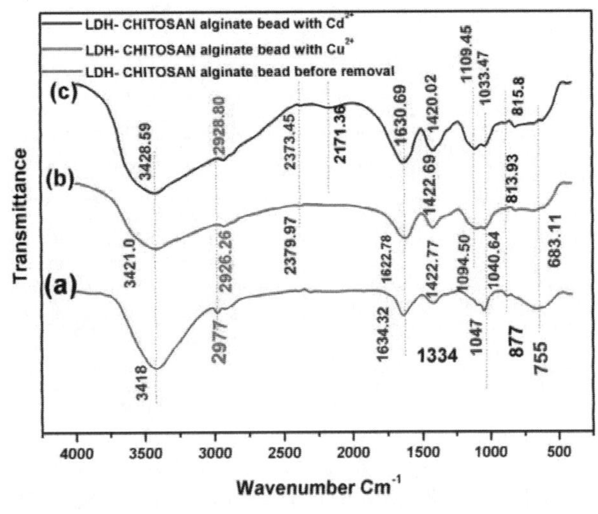

Figure 4. FT-IR spectra of Ni–Al LDH- chitosan – alginate beads (a) before adsorption, (b)after adsorption of cupper Cu^{2+} and (c) after adsorption of cadmium Cd^{2+} from different aqueous solutions.

4. Conclusion

Qarun Lake is an inland closed saline lake about 80 km southwest of Cairo. The unique source of the lake water is the agricultural wastewater of El Fayoum province. Due to the evaporation process that led to the concentration of the impounded drainage water that effect on the water quality and the environmental conditions of the lake. The status of the degraded lake water was confirmed with the results of WQI. The obtained WQI data classified Qarun Lake water as very poor water for fishing usage. On the other side, the metal pollution indices indicated that Qarun Lake suffers from different degree of metal contamination. Novel synthesized nanocomposites were used to eliminate the trace metals from the different wastewater that discharge into the lake.

The study showed that the iron is the most adsorbed one of the studied metals (Fe, Mn, Pb, Zn, Cu, Cd, Cr and Ni). Furthermore, the synthesized LDH-chitosan–alginate beads recorded the maximum removal efficiency > chitosan–alginate > LDH-alginate > alginate. It is worth mentioning that the synthesized nanocomposites considered an excellent adsorbent to remove different metals from the various wastewaters.

Table 11. FT-IR spectra assignment of Ni–Al LDH- chitosan – alginate beads (a) before adsorption, (b)after adsorption of cupper Cu^{2+} and (c) after adsorption of cadmium Cd^{2+} from different aqueous solutions.

(a) Ni-Al LDH /Chitosan Alginate beads before adsorption	(b) Ni-Al LDH/Chitosan Alginate beads with Cu^{2+}	(c) Ni-Al LDH/Chitosan Alginate beads with Cd^{2+}	Assignment	References
3418 cm^{-1}	Shift to 3421 cm^{-1}	Shift to 3428.6 cm^{-1}	stretching OH vibration mode	[46]
2977 cm^{-1}	Shift to 2926 cm^{-1}	Shift to 2928.6 cm^{-1}	Antisymmetric streching vibration mode of –CH_3	[47]
----------	2379 cm^{-1}	Shift to 2373 cm^{-1}	is calculated by υ(−C=O) of the amide group CONHR of the chitosan. & adsorption for cadmium and cupper.	[13,48]
----------	-----------	2171 cm^{-1}	For Cd^{2+}	[13]
1634 cm^{-1}	Shift to 1622 cm^{-1}	Shift to 1630 cm^{-1}	N-H bond stretching vibration &(NH_2 bending).	[49,50]
1422.7 cm^{-1}	Shift to 1422.6 cm^{-1}	Shift to 1420 cm^{-1}	Carboxyl group symmetric stretching vibration	[47]
	1094.5 cm^{-1}	1109 cm^{-1}	skeletal vibration including the COO stretching.	[50]
1047 cm^{-1}	Shift to 1040.6 cm^{-1}	Shift to 1033.5 cm^{-1}	-CH-OH in cyclic alcohols and C-O stretch	[51]
877 cm^{-1}	Shift to 813.8 cm^{-1}	Shift to 815.8 cm^{-1}	attributed to ω(C–H) from the polysaccharide's structure	[48]
755 cm^{-1}	Shift to 633 cm^{-1}	---------	Overlap between deformation of water molecules band in interlayer at 770 cm−1 and and at 680 cm−1 of carbonate ions & at 633 cm^{-1} ensue from metal–oxygen bonds M-O vibration in the brucite-like.	[27,52]

References

[1] Liu J, Dorjderem A, Fu J, Lei X and Macer D 2011 *Water Ethics and Water Resource Management.* (UNESCO)

[2] Abdel-Satar A M, Goher M E and Sayed M F 2010 Recent environmental changes in water and sediment quality of Lake Qarun, Egypt *J. Fish. Aquat. Sci.* **5** 56–69

[3] Dardir A A and Wali A M A 2009 Extraction of salts from lake Quaroun, Egypt: Environmental and economic impacts *Glob. Nest J.* **11** 106–13

[4] Reemtsma T and Jekel M 2006 *Organic pollutants in the water cycle: properties, occurrence, analysis and environmental relevance of polar compounds* (John Wiley & Sons)

[5] Doherty V, Ogunkuade O and Kanife U 2010 Biomarkers of oxidative stress and heavy metal

levels as indicators of environmental pollution in some selected fishes in Lagos, Nigeria *Am. J. Agric. Environ. Sci.* **7** 359–65

[6] Liu Y, Liu J, Zhang A and Liu Z 2017 Treatment effects and genotoxicity relevance of the toxic organic pollutants in semi-coking wastewater by combined treatment process *Environ. Pollut.* **220** 13–9

[7] El-Sayed M and Salem W M 2015 Hydrochemical assessments of surface Nile water and ground water in an industry area – South West Cairo *Egypt. J. Pet.* **24** 277–88

[8] Farombi E O, Adelowo O A and Ajimoko Y R 2007 Biomarkers of oxidative stress and heavy metal levels as indicators of environmental pollution in African cat fish (Clarias gariepinus) from Nigeria Ogun River *Int. J. Environ. Res. Public Health* **4** 158–65

[9] Goher M E, Hassan A M, Abdel-Moniem I A, Fahmy A H and El-sayed S M 2014 Evaluation of surface water quality and heavy metal indices of Ismailia Canal, Nile River, Egypt *Egypt. J. Aquat. Res.* **40** 225–33

[10] Tesh S J and Scott T B 2014 Nano Composites for Water Remediation: A Review *Adv. Mater.* **26** 6056–68

[11] Zhang Y, Wu B, Xu H, Liu H, Wang M, He Y and Pan B 2016 Nanomaterials-enabled water and wastewater treatment *NanoImpact* **3** 22–39

[12] Moaty S A A, Farghali A A, Moussa M and Khaled R 2017 Remediation of waste water by Co–Fe layered double hydroxide and its catalytic activity *J. Taiwan Inst. Chem. Eng.* **71** 441–53

[13] Mahmoud R, Moaty S A, Mohamed F and Farghali A 2017 Comparative Study of Single and Multiple Pollutants System Using Ti–Fe Chitosan LDH Adsorbent with High Performance in Wastewater Treatment *J. Chem. Eng. Data* **62** 3703–22

[14] Moaty S A A, Farghali A A and Khaled R 2016 Preparation, characterization and antimicrobial applications of Zn-Fe LDH against MRSA *Mater. Sci. Eng. C* **68** 184–93

[15] Shadrin N V, El-Shabrawy G M, Anufriieva E V, Goher M E and Ragab E 2016 Long-term changes of physicochemical parameters and benthos in Lake Qarun (Egypt): Can we make a correct forecast of ecosystem future? *Knowl. Manag. Aquat. Ecosyst.* 18

[16] Ibrahim L A and Ramzy E M 2013 Water quality and its impact on Tilapia zilli (case study) Qarun lake-Egypt *Int. Water Technol. J.* **3** 170–91

[17] El-Shabrawy G M, Anufriieva E V, Germoush M O, Goher M E and Shadrin N V 2015 Does salinity change determine zooplankton variability in the saline Qarun Lake (Egypt)? *Chinese J. Oceanol. Limnol.* **33** 1368–77

[18] Cude C G 2001 Oregon Water Quality Index a Tool for Evaluating Water Quality Management Effectiveness 1 *JAWRA J. Am. Water Resour. Assoc.* **37** 125–37

[19] Caeiro S, Costa M H, Ramos T B, Fernandes F, Silveira N, Coimbra A, Medeiros G and Painho M 2005 Assessing heavy metal contamination in Sado Estuary sediment: an index analysis approach *Ecol. Indic.* **5** 151–69

[20] Bakan G, Özkoç H B, Tülek S and Cüce H 2010 Integrated Environmental Quality Assessment of K ı z ı l ı rmak River and its Coastal Environment *Turkish J. Fish. Aquat. Sci.* **462** 453–62

[21] Tamasi G and Cini R 2004 Heavy metals in drinking waters from Mount Amiata (Tuscany, Italy). Possible risks from arsenic for public health in the Province of Siena *Sci. Total Environ.* **327** 41–51

[22] El-Rouby W M A., Goher M E., Noaemy S G. E-D S I Efficient Water decontamination Using Nanocomposite of Layer Double Hydroxide Beads.*Under puplication*

[23] Y A A T *Drains impact on water quality of manzala lake and photocatalytic removal of some chemical pollutants via using modified graphene nanomaterials* (B.Sc.Thesis, Fac. of Sci. Benha University)

[24] E G M*Chemical studies on the precipitation and dissolution of some chemical elements in Lake Qarun.* (PDh Thesis. Fac. Sci. Al-Azhar Univ.Cairo, Egypt.)

[25] Khalil M T, Fishar M R, Shakir S H, Amer A S and Nassif M G Impact of Drainage Water on

Macrobenthos Structure of Lake Qaroun, El-Fayoum, Egypt
[26] Fishar M R, El-Mageed A A, Elwa S E A-E A S and B R E . Environmental factors affecting life cycle of benthic invertebrates in Lake Qaroun, Egypt. *J.Egypt. Acad. Soc. Environ.Develop. (D- Environ. Stud.* **6** 932
[27] Ahmed A A A, Talib Z A, Bin Hussein M Z and Zakaria A 2012 Zn-Al layered double hydroxide prepared at different molar ratios: Preparation, characterization, optical and dielectric properties *J. Solid State Chem.* **191** 271–8
[28] Rostom N G, Shalaby A A, Issa Y M and Afifi A A 2017 Evaluation of Mariut Lake water quality using Hyperspectral Remote Sensing and laboratory works *Egypt. J. Remote Sens. Sp. Sci.* **20** S39–48
[29] Saffran K, Cash K, Hallard K, Neary B and Wright R 2001 Canadian water quality guidelines for the protection of aquatic life *CCME water Qual. Index* **1** 31–4
[30] E-S M H K 2010 *Quantitative assessment and treatment of some industrial pollutants along the river Nile at Giza region* (Ph.D. Thesis, Fac. of Sci. Benha University)
[31] Goher M E, Mahdy E-S M, Abdo M H, Farida M, Korium M A and Elsherif A A S Water quality status and pollution indices of Wadi El-Rayan lakes, El-Fayoum, Egypt *Sustain. Water Resour. Manag.* 1–14
[32] Kavita U, Piyush M and Gupta A K 2010 Studies on the physico-chemical status of two ponds at Varanasi and Bhadohi under biotic stress. *Plant Arch.* **10** 691–3
[33] Edwards A C and Withers P J A 2008 Transport and delivery of suspended solids, nitrogen and phosphorus from various sources to freshwaters in the UK *J. Hydrol.* **350** 144–53
[34] Mueller D K, Helsel D R and Kidd M A 1996 *Nutrients in the nation's waters: too much of a good thing?* (US Government Printing Office Washington, DC)
[35] Sahni K and Yadav S 2012 Seasonal variations in physico-chemical parameters of Bharawas Pond, Rewari, Haryana *Asian J. Exp. Sci* **26** 61–4
[36] Abdel-Satar A M, Ali M H and Goher M E 2017 Indices of water quality and metal pollution of Nile River, Egypt *Egypt. J. Aquat. Res.* **43** 21–9
[37] Bahnasawy M, KHIDR A A and Dheina N 2011 Assessment of heavy metal concentrations in water, plankton, and fish of Lake Manzala, Egypt *Turkish J. Zool.* **35** 271–80
[38] Goher M E, Hassan A M, Abdel-Moniem I A, Fahmy A H, Abdo M H and El-sayed S M 2015 Removal of aluminum, iron and manganese ions from industrial wastes using granular activated carbon and Amberlite IR-120H *Egypt. J. Aquat. Res.* **41** 155–64
[39] Khang H V, Hatayama M and Inoue C 2012 Arsenic accumulation by aquatic macrophyte coontail (Ceratophyllum demersum L.) exposed to arsenite, and the effect of iron on the uptake of arsenite and arsenate *Environ. Exp. Bot.* **83** 47–52
[40] Gheju M 2011 Hexavalent chromium reduction with zero-valent iron (ZVI) in aquatic systems *Water, Air, Soil Pollut.* **222** 103–48
[41] Ruttner F 1963 *Fundamentals of limnology* (University of Toronto Press)
[42] Sarkar C and Abbasi S A 2006 Qualidex--a new software for generating water quality indices. *Environ. Monit. Assess.* **119** 201–31
[43] Federation W E and Association A P H 2005 Standard methods for the examination of water and wastewater *Am. Public Heal. Assoc. Washington, DC, USA*
[44] Rahmanian O, Maleki M H and Dinari M 2017 Ultrasonically assisted solvothermal synthesis of novel Ni/Al layered double hydroxide for capturing of Cd (II) from contaminated water *J. Phys. Chem. Solids* **110** 195–201
[45] Asiabi H, Yamini Y and Shamsayei M 2017 Highly selective and efficient removal of arsenic (V), chromium (VI) and selenium (VI) oxyanions by layered double hydroxide intercalated with zwitterionic glycine *J. Hazard. Mater.* **339** 239–47
[46] Vijayalakshmi K, Devi B M, Latha S, Gomathi T, Sudha P N, Venkatesan J and Anil S 2017 Batch adsorption and desorption studies on the removal of lead (II) from aqueous solution using nanochitosan/sodium alginate/microcrystalline cellulose beads *Int. J. Biol. Macromol.* **104** 1483–94

[47] Ruan X, Chen Y, Chen H, Qian G and Frost R L 2016 Sorption behavior of methyl orange from aqueous solution on organic matter and reduced graphene oxides modified Ni-Cr layered double hydroxides *Chem. Eng. J.* **297** 295–303

[48] Negrea P, Caunii A, Sarac I and Butnariu M 2015 THE STUDY OF INFRARED SPECTRUM OF CHITIN AND CHITOSAN EXTRACT AS POTENTIAL SOURCES OF BIOMASS. *Dig. J. Nanomater. Biostructures* **10**

[49] Kumar N, Reddy L, Parashar V and Ngila J C 2017 Controlled synthesis of microsheets of ZnAl layered double hydroxides hexagonal nanoplates for efficient removal of Cr (VI) ions and anionic dye from water *J. Environ. Chem. Eng.* **5** 1718–31

[50] Peniche C, Argüelles-Monal W, Davidenko N, Sastre R, Gallardo A and San Román J 1999 Self-curing membranes of chitosan/PAA IPNs obtained by radical polymerization: Preparation, characterization and interpolymer complexation *Biomaterials* **20** 1869–78

[51] Nayak P L and Sahoo D 2011 Chitosan-alginate composites blended with cloisite 30B as a novel drug delivery system for anticancer drug paclitaxel *Int. J. Plast. Technol.* **15** 68–81

[52] Jitianu M, Gunness D C, Aboagye D E, Zaharescu M and Jitianu A 2013 Nanosized Ni-Al layered double hydroxides Structural characterization *Mater. Res. Bull.* **48** 1864–73

IOP Conf. Series: Materials Science and Engineering **464** (2018) 012004 doi:10.1088/1757-899X/464/1/012004

Extraction of degradable bio polymer materials from shrimp shell wastes by two different methods

M Faisal[1*], A Elhussieny[1], K A Ali [2], I Samy [3]and N M Everitt [4]

[1]Centre of Nano-science and technology (CNT), The Nile University, Nile Avenue, Giza, 116453, Egypt.
[2]Centre of Excellent for advanced science, nanotechnology group, applied Organic chemistry department, National Research centre, Dokki, Giza 12622-Egypt.
[3]Department of Industrial Engineering, School of Engineering, The Nile University, Nile Avenue, Giza, 116453 Egypt.
[4]Department of Mechanical Materials and Manufacturing Engineering, University of Nottingham, Nottingham, NG7 2RD, UK.

* Corresponding author, e-mail: m.faisal@nu.edu.eg

Abstract. Chitosan is one of the most widespread biopolymer materials available in nature which is extracted from chitin. The main source of chitosan is the chitin that is extracted from exoskeletons of crustacean, such as shrimp and crabs, which are found in a huge amount of shells waste that produced from seafood companies around the world. The chitosan has several applications such as pharmaceutical, fertilizer and edible coating in food industries. The quality of the chitosan's depends on its extraction method, so in this research work we have studied the effect of grounding shrimp shell waste before the extraction step. The chemical and physical properties of the extracted chitosan were studied by using different techniques such as Fourier transform infrared spectroscopy (FTIR), and Thermogravimetric analysis (TGA). The study showed that the films and powder produced from non-ground samples showed a higher purity of chitosan and better thermal stability than the chitosan produced from ground samples.

1. Introduction

Chitosan is a natural biopolymer material produced from chitin biopolymer, which is the second natural abundant biopolymer found in nature after cellulose. Chitin has been found in many sources such as marine crustaceans, insects, some algae, and fungi [1]. Chitosan is non- toxic, biodegradable and antimicrobial material which has many beneficial applications in many fields such as food industry, water treatment, manufacturing of membranes, cosmetics, antifungal and antibacterial products [2]. Chitosan is a bi-product extracted from chitin by using alkali treatment which involves three main steps. The first step is deproteinization (DP) step, which removes proteins and undesired materials such as pigments from shrimp shells. This is followed by demineralization (DM), which removes all minerals (e.g. $CaCO_3$) by using acidic solution, then the deacetylation step (DD) which removes acetyl groups from the polymeric chain. The produced chitosan's quality depends on the conditions of extraction such as alkali concentrations and reaction time of acid with samples [3].

In this study, we explored the effect of the mechanical treatment of the shrimp shell waste on the yield of chitosan extracted. Degradable biopolymer films were made from chitosan using two different methods of extraction. The first method was based on grounding shrimp shell waste as shown in figure 1a, whilst the second method relied on treatment of shrimp shells without groundings shown in Figure 1b. The research tested the hypothesis that grounding the shells increases the surface area and would thus affect the properties of the chitosan produced. Various procedures have been adopted to obtain chitosan from crustaceans. Although chemical methods are widely used to extract chitosan and chitin from the raw materials shrimp wastes [4], there is little information about how to make the production process more efficient. Chitosan extraction usually takes place by enzymatic or chemical methods. Demineralization usually takes place by HCl and deproteinization by NaOH. The sequence of these extraction steps differs, although in most chitin producing industries the deproteinization step is carried out before demineralization [5].

Figure 1. a) ground shrimp, b) whole shrimp without grounding.

Divya. K, et al reported a higher w/w % yield of high purity chitosan and the difference in yield was possibly due to the increased time of reaction [6]. Deproteinization and demineralization steps were repeated twice and, the final deacetylation of chitin at room temperature lasted 3 days. The characteristics of produced chitosan were in accordance with the commercial standard. The obtained chitosan had low viscosity, high degree of deacetylation (DD) and a denser crystalline structure. Chitosan with such properties have many commercial applications and greater scope of industrial applications.

M. S. Hossain and A. Iqbal reported chitosan produced from shrimp waste by a chemical method that started with the demineralization step before deproteinization and deacetylation steps [7]. Carrying out demineralization with different concentrations of HCl and NaOH in deproteinization steps, the results showed that 3% HCl used in demineralization and 4% NaOH in the deproteinization were the suitable concentration for these methods, respectively, at ambient temperature (28±2°C)without a heating process. The degree of deacetylation (DD) process depends on NaOH concentration. Due to the fact that the highly bonded acetyl groups in chitin is difficult to remove, higher concentration of NaOH and higher temperature are required. In this case, the increase of NaOH concentration improved the deacetylation process where the highest deacetylation degree (81.24%) was reached at a NaOH concentration of 60%. The improvement in deacetylation degree was not observed with NaOH concentration more than 60% since the process become inefficient especially in the final washing to obtain chitosan. Among the treatments used in the study 4% NaOH and 3% HCl were found to most successfully extract chitin. The highest deacetylate yield of chitosan with maximum solubility was done by 60% NaOH treatment, whilst high quality chitosan DD 79.57% and 97.02% solubility with minimum chemical usage was yielded by 50% NaOH treatment.

N.S. Mahmoud et al. [8] showed the effect of using organic acids in the demineralization on the deproteinized chitin instead of using inorganic acid HCl. The study showed that the effectiveness of using lactic and/or acetic acids for demineralization of shrimp shells was comparable to that of using hydrochloric acid. Since organic acids are environmental friendly and the chitin properties are maintained, the advantages of using organic acids in the demineralization step would result in: (a) production of important products as acetate and lactate salts besides chitin production. (b) Successful removing of minerals, (c) maintaining natural chitin properties and (d) reduction in the purification cost [9], however the conventional HCl was used in this study.

2. Experimental section

2.1. Materials

The raw shrimp shell waste was bought from local fish market, Al-Obour city, Egypt. The shrimp shell waste was obtained from mix of species of shrimp. All of the reagents used were of a high analytical grade which purchased from Sigma-Aldrich chemical company; Sodium Hydroxide pellets (NaOH, pure), Hydrochloric acid (HCl, 35-38%), Acetic Acid (CH_3COOH, 100%).

2.2. Methods of extraction

Approximately 2000 g of shrimp shells waste was washed several times with tap water then deionized water until the shell waste were clean and then the shells were dried at room temperature. The dried shrimp shell waste was divided into two samples (A) and (B) each weighing ~1000 g, (A) was ground with an electric grinder and (B) was left as the whole exoskeleton with head and tail without grinding. The following procedures were applied for both (A) and (B). Chitin was extracted from shrimp shells by deproteinized followed by demineralization steps [10]. Shells were deproteinized by boiling several times in 1 M NaOH solution and demineralized using 1 M HCl. Then chitin produced was deacetylation using 50% NaOH to obtain chitosan.

2.3. Preparation of chitosan film by casting

For each sample, 2g of chitosan was added to 1 % CH_3COOH then boiled for 30 mins, and then poured into flat flexible plastic cover [11]. The samples were left to dry at room temperature, until solid films were produced as shown in figure 2.

2.4 Characterization Tests

2.4.1 pH and Solubility. The pH measurement of the chitosan solutions was carried out using a pH meter (EC500). Solubility of the chitosan was tested by dispersing 0.5 g of chitosan in 1% acetic acid and heating the mixture to 100 ^{0}C with continuous stirring. If the formed chitosan was not soluble then the degree of deacetylation would not be considered acceptable and the deacetylation process would need to be repeated

2.4.2 Fourier transformation Infrared (FT-IR) Spectra. FT-IR spectral analysis of chitosan is used to detect the functional groups of chitin and chitosan and to determine the degree of deacetylation DD, but it is valid only when the value of DD is not more than 55 % [12]. The spectral region was scanned between 4000 and 400 cm-1. Potassium bromide KBr pellets were used in preparing specimens. Dried powdery chitosan was mixed with KBr powder and pressed in vacuo to form a homogeneous disc with a thickness of 0.5 mm. The chitin concentration in the samples was 2%, calculated with respect to KBr.

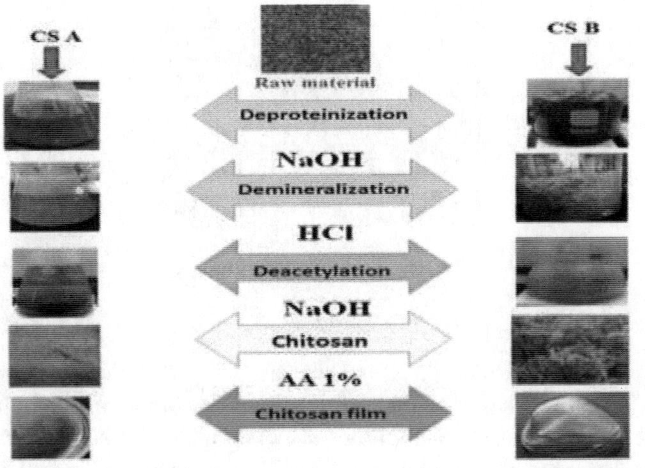

Figure 2. The extraction steps of chitosan and preparation chitosan film from method (A) and method (B).

2.4.3 ¹HNMR spectroscopic analysis. 1HNMR spectroscopy is used for determining DD of chitin/chitosan using the following equation [13]:

DD%= [1/3X ICH3 ÷1/6 X I(H2-H6)]X100

The chitosan was dissolved in DMSO/trifloroacetic acid and the spectroscopic analysis carried out on a Bruker Avance III instrument (500 MHz). Nomenclature is given as recommended by Harris et al [13]. Chemical shifts (δ) are given in ppm and coupling constants J in Hz. Relax delay 1.00 sec. pulse 45.0 degree. The chitosan samples were with concentration between 0 and 10 ppm.

2.4.4 Thermal Gravimetric Analysis (TGA). Thermal stability of chitosan powders and films was measured by Thermal gravimetric analysis (TGA) Using a Hi-Res TGA 2950 Thermogravimetric Analyzer from TA Instruments Company (Tokyo, Japan. The analyses were made by increasing the temperature from room temperature to 800°C in an inert nitrogen atmosphere with a flow rate of 60 mL/min and warming rate of 10 °C/min

2.4.5 Zeta potential. Zeta potential of the particles was recorded on a Particle Sizing Systems\ZPW388-V2.14\zpw388.tbl.The wavelength used was 632.8 nm, and sample temperature of 230°C

2.4.6 Measurement of Viscosity. Viscosity was measured using Brookfield viscometer 1% chitosan solution was prepared using 1% acetic acid. Measurement was made using a No. 5 spindle at 50 rpm at 25 °C with values reported in centipoises (cPs) units.

2.4.7 Calculation of Yield Percentage. The amount of chitosan extracted from 1000g of dry shrimp shell wastes was calculated as yield percentage as following:
Yield %= (Mass of dry chitosan extracted/mass of dry shell waste) x100.

3. Results and Discussion

3.1. pH and Solubility test

The pH of dissolved chitosan from method A was around 11- 12, while the pH from (B) was about 9-10. When the solubility test was conducted Both extracted methods produced chitosan which dissolved, as shown in figure 3 but chitosan from method A took a longer time to dissolve than from method B.

IOP Conf. Series: Materials Science and Engineering **464** (2018) 012004 doi:10.1088/1757-899X/464/1/012004

Figure 3. Chitosan solution.

3.2. Fourier transformation Infrared (FT-IR) Spectra

FTIR analysis was conducted to observe structural changes that occurred as a result of chitosan produced by two different methods. It was based on the identification of the functional groups present in chitosan molecules, which are given in Table 1 with the corresponding absorption bands (in cm-1). In order to fully characterize the starting materials, chitosan A and B were compared with the spectra of pure high molecular weight chitosan (HMWC) and low molecular weight chitosan (LMWC). The first step in the procedure is the conversion of chitin into chitosan, which was confirmed through the disappearance or decrease of peaks that are related to secondary amides (thus to chitin) in both experimental procedures. The FT-IR spectra of the two chitosan samples produced from methods A and B are shown in figure 4 and revealed that there is a reduction in the band at 1629 cm^{-1} in CsA (Chitosan Method A), as it transformed from chitin to chitosan. In the pure LMWC and HMWC samples no band was observed between 1650 and 1900 cm^{-1}, which showed that oxidative groups such as carboxylic aldehyde, or carbonyl groups did not exist in these samples.

But in CsA a new absorption ban showed at 1797.5 cm^{-1} which is attributed to the absorption of the carboxylic group (–COOH). These results suggest the rupture of β-(1, 4) glycoside linkage in macromolecule of CsA. Furthermore, significant decreases in the intensity of 1629 cm^{-1} showed the effective of the deacetylation process. The FT-IR spectrum of chitosan from method B (CsB) showed the characteristic absorption bands at 3442.31, 2921.63 cm^{-1} corresponding to NH and CH stretching. The bands around 1643.05,1589.06 and 1421.28 cm^{-1} were assigned to amide I, II and III, respectively, while the characteristic bands attributed to the saccharide structure appeared at 1079.94 cm-1(anti-symmetric stretching of the C-O-C bridge). Absorption band at 1643.05cm^{-1} assigned to the increase of deacetylation degree following with the same increase of intensity at 1589.06 cm^{-1} which shows free amine group NH_2, as shown in table 1.

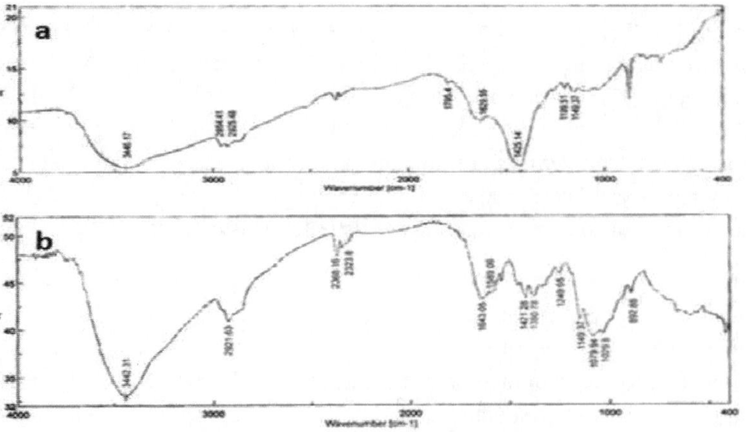

Figure 4. FTIR for a) method A, b) for method B.

Table 1. Comparison between FTIR spectral Patterns between commercial and extracted chitosan

Commercial Chitosan		**Experimental Chitosan**		**Vibration Modes**
High MWT	**Low MWT**	**Cs *Meth.A***	**WSCs *Meth.B***	
3446.17	3432.67	3446.17	3442.31	NH_2,OH in pyranose ring
2923.56	2921.63	2954.41 , 2925.48	2921.63	CH2 in CH2OH group
2360.44	2362.37		2368.16	
2333.45	2339.23	1795.4	2323.8	(C=O)
1646.91	1652.7	1629	1643.05 1589.06*	(C=O) in NHCOCH3 group (Amide I band) *NH_2in amino group
---------	1421.23	1425.14	1421.28	CH_3
1396.21	1378.85	--------	1380.78	CH2 in CH_2OH group
________	1321 (-CH3) tertiary amide	--------	1249.65 (OH) or C–O-C Bands	(C-H) in pyranose ring
1149.37	1153.22		1149.37	–C-H complex vibrations of NHCO group (Amide III band)
1076.08	1076.08	----------	1079.94	(C-O-C) (glyosidic linkage)
---------	896.737	______	892.88	pyranose ring skeletal vibrations

---------- no band found

3.3.^{1}HNMR

The ^{1}HNMR spectra are accurate and precise for measuring high degree of deacetylation (DDA), which is usually difficult to measure using other techniques like IR or titration of dilute solutions of Chitosan samples dissolved in DMSO solvent.
The ^{1}HNMR spectrum of method A, revealed peaks at δ=4.79 (H1), δ =2. (H2), δ =3.43~3.686 (H3, H4, H5, H6), δ =2.1 (NHCOCH3) ppm, while in method B; the 1HNMR spectrum revealed peaks at the same positions but with lower intensities, as shown in Figure 5. The Peaks appeared at 2.5, 3.5 ppm related to the solvent. The DD of method A and Method B results are shown in Table 2. The higher DD for method B ties in with the greater solubility reported found in the solubility test

Table 2. The DD of method A and Method B

Method A	**57%**
Method B	**70%**

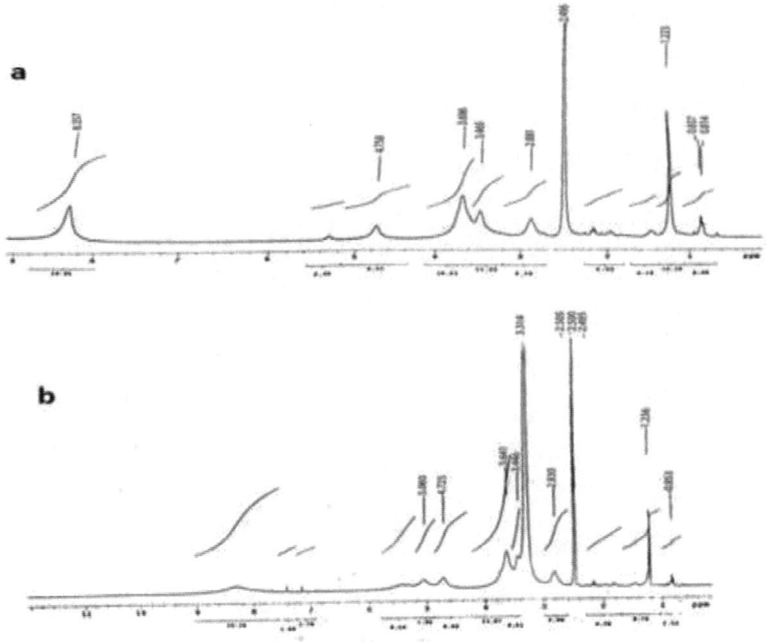

Figure 5. ^{1}HNMR for a) method A, b) for method B.

3.4. Zeta potential

The magnitude of zeta potentials as seen in Figure 6, shows method B averaged 48.98, while the average for method A was 37.88. The higher magnitude of zeta potential in method B. Themagnitude of zeta potential reflects the degree of electrostatic repulsion between charged particles. For dispersion system, high degree of zeta potential, negative or positive indicates resistance of the particles to aggregation, and this denotes an apparently stable system [14].

3.5. Thermogravimetric (TGA)

The thermal stability of chitosan particles was investigated by thermogravimetric analysis. The changes in sample weight with the increase of the temperature are presented in Figure 7. Chitosan weight loss curves associated with the decomposition of chitosan are in the range 200–450 0C15. In our study, there is a weight loss up to 100 ◦C in all samples due to elimination of adsorbed water that shows hydrophobic character. Amongst all the samples, method B the non-ground chitosan has more residual weight after TGA and is thus considered to be the most stable chitosan in this study. Samples showed a significant weight loss from 270 ^{0}C up to 500 ^{0}C. This can be attributed to the pyrolytic decomposition, dehydration of the saccharide rings and depolymerisation of the polysaccharide structure16. Analysed the gases evolved during TGA using FTIR, the complex gaseous mixture released during the degradation of chitosan was mainly composed of CO_2, NH_3, CO.CH_3COOH and CH_4. The release of H_2O, NH_3, CO_2, CO and CH_3COOH was observed in the temperature range 2500 ˚C - 450 ˚C and this assigned to the pyrolytic degradation of chitosan.NH_3 was released at lower temperature than the other gases. While the release of CH_4 took place in the range from 450 ˚C -750 ˚C. As suggested in literature17-19, this step suggests a modification of the material when the structure is decreased due the release of methane and forms graphite-like structures through a hydrogenation mechanism.

Figure 6 . Zeta potential a) for method A, b) for method B

According to the TGA rate of wt% loss curve of chitosan-CsA (blue line in shown in Figure 7a (a), there are multi-degradation stages. In the first stage, weight loss starts at ~<100°C and continues, reaching at the second stage its maximum weight loss (77. %/°C) at 307 °C and at 355 °C another peak where it lost 63%/°C. The last degradation stage curve at 639°C with a weight loss percentage of 30%/°C which indicates the elimination of CO_2 gas. The chitosan from method (A) film also showed similar multi degradation steps starting at less than 100°C up to 700°C reaching its maximum weight loss 79 %/°C at 290°C as shown in Figure 7a (b). In the rate of weight loss curve of chitosan from method (B) as shown in Figure 7b (c), there two degradation stages reaching its maximum weight loss 66%/°C at 307°C. The films produced from Method B are illustrated in Figure 7b(d) and revealed three degradation stages. First degradation stage below 100°C, maximum rate weight loss at 283°C with highest percentage 67%/°C, and then last degradation stage at 468°C with weight loss 42.17%/°C. The residues of the four samples were recorded as shown in table 3.

Table 3. Residues of the four samples

Samples residue	weight(mg)
Chitosan A	0.3127
Chitosan film A	0.5191
Chitosan B	0.6209
Chitosan B film	1.187

3.6. Measurement of viscosity

The viscosity of chitosan A (η) was found to be 24 cP, whilst Chitosan B showed a viscosity (η) of 180 cP. Hence method B, produced chitosan with higher viscosity.

3.7. Calculation of Yield Percentage

Amount of Chitosan extracted from the ground waste method was 187 g from 1000g of shell wastes, whilst chitosan from the whole exoskeleton method produced 226.2 g from1000 g of shell wastes. Hence the yield of CsA was lower (18.7wt. %) than that of CsB (22.62wt. %).

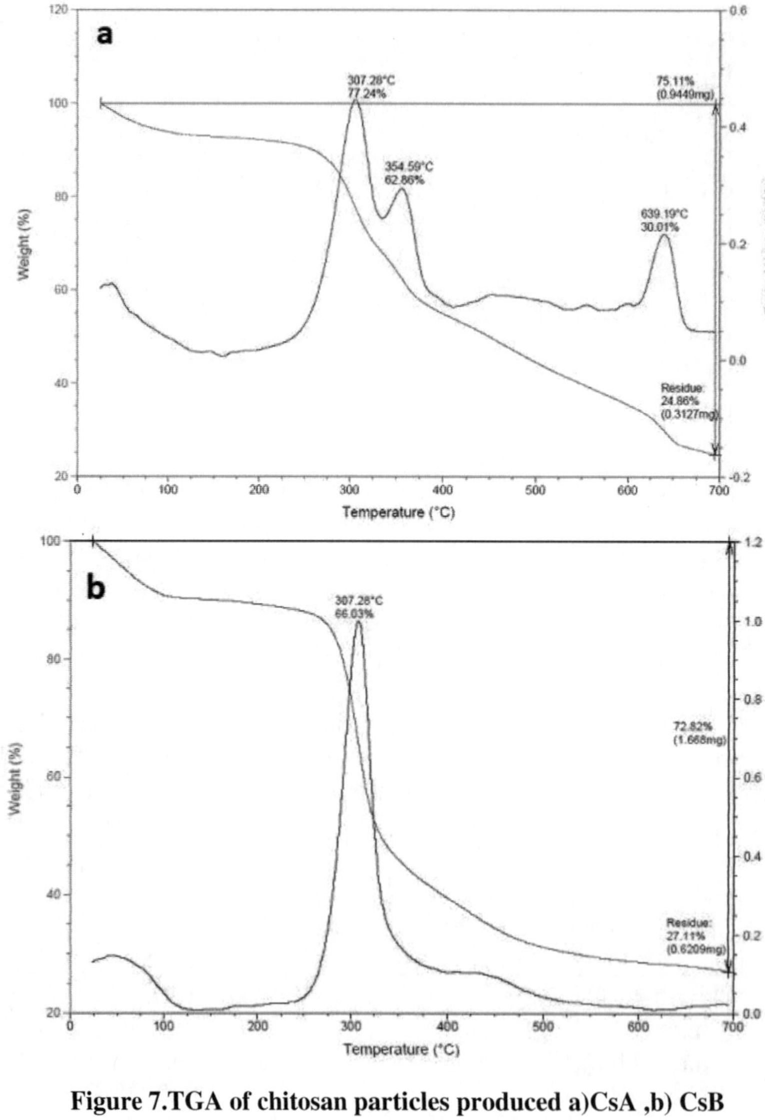

Figure 7.TGA of chitosan particles produced a)CsA ,b) CsB

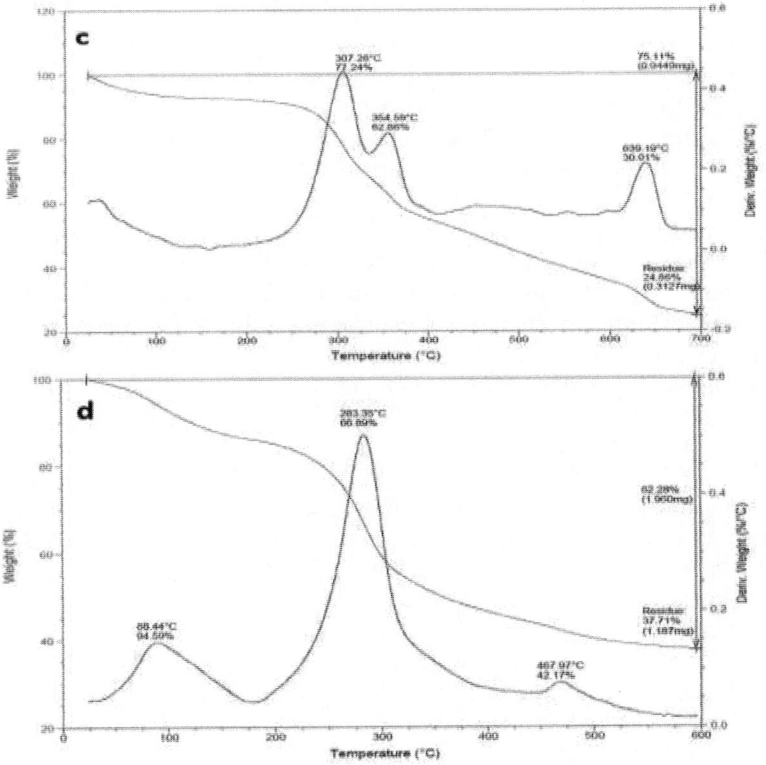

Figure7b.TGA of chitosan films produced from methods c)Cs A, d)CsB

4. Conclusion

Chitosan produced from whole, non-ground dried exoskeleton (method B) showed a higher degree of deacetylation, a higher purity of chitosan, and enhanced thermal stability compared to chitosan from produced from ground dried shrimp waste (method A). Using shrimp shell waste as a whole shell showed a higher degree of control of the extraction procedure. With the chemical treatment described in this paper, a larger yield of chitosan was produced from non- ground than ground shrimp shell method.

Acknowledgment

The authors kindly thank Newton-Mosharfa fund, Science &Technology development Fund in Egypt (STDF), British council for funding this project.

References

[1] Sonawane S, Setty Y P and Sapavatu S N 2015 *Chemical and bioprocess engineering: Trends and developments* (CRC Press)

[2] Gómez-Ríos D, Barrera-Zapata R and Ríos-Estepa R 2017 Comparison of process technologies for chitosan production from shrimp shell waste: A techno-economic approach using Aspen Plus® *Food Bioprod. Process.* **103** 49–57

[3] Nouri M, Khodaiyan F, Razavi S H and Mousavi M 2016 Improvement of chitosan production from Persian Gulf shrimp waste by response surface methodology *Food Hydrocoll.* **59** 50–8

[4] Domard A, Roberts G A F and Varum K M 1997 Advances in chitin science *Domard, A* 410

[5] Muzzarelli R A A, Peter M G and European Chitin Society. 1997 *Chitin handbook* (Atec)

[6] Divya K, Sharrel R and Jisha S M 2014 A simple and effective method for extraction of high purity chitosan from shrimp shell waste *Proceedings of the international conference on advances in applied science and environmental engineering-ASEE*

[7] Hossain M S and Iqbal A 2014 Production and characterization of chitosan from shrimp waste *J. Bangladesh Agril. Univ* **12** 153–60

[8] Mahmoud N S, Ghaly A E and Arab F 2007 Unconventional approach for demineralization of deproteinized crustacean shells for chitin production *Am. J. Biochem. Biotechnol.* **3** 1–9

[9] Fahim I S, Marei N, Salem H G and Mamdouh W 2015 Effect of graphene and fullerene nanofillers on controlling the pore size and physicochemical properties of chitosan nanocomposite mesoporous membranes *J. Nanomater.* **16** 114

[10] Aryaei A, Jayatissa A H and Jayasuriya A C 2012 Nano and micro mechanical properties of uncross-linked and cross-linked chitosan films *J. Mech. Behav. Biomed. Mater.* **5** 82–9

[11] Paul S, Jayan A, Sasikumar C S and Cherian S M 2014 Extraction and purification of chitosan from chitin isolated from sea prawn Fenneropenaeus indicus *EXTRACTION* **7**

[12] Baxter A, Dillon M, Taylor K D A and Roberts G A F 1992 Improved method for ir determination of the degree of N-acetylation of chitosan *Int. J. Biol. Macromol.* **14** 166–9

[13] Harris R K, Becker E D, De Menezes S M C, Goodfellow R and Granger P 2001 NMR nomenclature. Nuclear spin properties and conventions for chemical shifts (IUPAC Recommendations 2001) *Pure Appl. Chem.* **73** 1795–818

[14] Rodrigues S, da Costa A M R and Grenha A 2012 Chitosan/carrageenan nanoparticles: Effect of cross-linking with tripolyphosphate and charge ratios *Carbohydr. Polym.* **89** 282–9

IOP Conf. Series: Materials Science and Engineering **464** (2018) 012005 doi:10.1088/1757-899X/464/1/012005

Bimodal applications of LDH-chitosan nanocomposite: water treatment and antimicrobial activity

W Kamal[1], Waleed M A El Rouby [2], Ahmed O El-Gendy[3], A A Farghali[2]*

[1] Department of chemistry, Faculty of Science, Beni- Suef University (BSU), Beni-Suef, Egypt.
[2] Material Science and Nano technology Department , Faculty of Postgraduate Studies for Advanced Sciences (PSAS), Beni-Suef University (BSU), Beni-Suef, Egypt.
[3] Microbiology and immunology department, Faculty of pharmacy, Beni-Suef University (BSU), Beni-Suef, Egypt.
* Corresponding author, e-mail: d_ farghali@yahoo.com

Abstract. Chitosan-LDH (layered doubled hydroxide) composite was prepared for removal heavy metals and antimicrobial activity. It was characterized by Fourier transform infrared spectrometer (FTIR), X-ray diffraction (XRD), Transmission electron microscopy (TEM), Scanning electronic microscope (SEM). Studying various physicochemical parameters such as the effects of pH, adsorbent amount, and contact time for removing heavy metals were performed. The adsorption kinetics and adsorption isotherm model were done to understand the adsorption process. The prepared composite exhibited strong antimicrobial activity against different gram positive (*Lactobacillus sakei* LMG 2313, *Corynebacterium sp*, *Staphylococcus aureus* 5247) and gram negative bacteria (*Escherichia coli AO5*). Results were recorded in terms of minimum inhibitory concentration (MIC), which is the lowest concentration of antimicrobial/antifungal agent that will inhibit the visible growth of a microorganism after overnight incubation.

Key Words: Chitosan, LDH, nanocomposite, adsorption, copper, iron, antimicrobial activity, waste water treatment.

1. Introduction

Water is the lifeblood and the most important component in the living cell (50-60% of the weight of the cell) [1]. Many human activities in the environment cause pollution of water due to production of chemical and physical contaminants [2]. Sources of water pollution include industrial waste, metal cleaning, electronic devices manufacturing, mining, fertilizer,... etc [2,3]. Toxic elements such as iron, copper, lead, nickel, mercury, chromium, cobalt and cadmium at high concentrations are the main source of industrial waste water and cause serious concern because of its negative effects on the environment, health and society [4,5]. These toxic elements deposit in the soil and reach to the plant and then to the animal and human causing many diseases [6]. For instance Arsenic (III) and (V), chromium (V1), Cadmium, Copper are carcinogens, whilst Mn (II), Fe (II) and (III) are toxics at high levels [7]. Water treatment and removal of heavy metals were performed by using different ways of chemical precipitation, coagulation, ion exchange, membrane separation, electro deposition and other methods [8]. All of these methods have high efficiency and good results in removal, but the high cost limit its use [1,4,9]. On the other hand, adsorption is one of the most common physical or chemical methods used for water treatment due to its use with small concentrations of heavy metals [1,4,5,9] in addition to cost-effective, simplicity and rapidness of operation, minimal sludge production, high efficiency and reusability.

Water pollution problems include presence of many microorganisms such as bacteria colon group, which is the main source of intestinal diseases, and Salmonella bacteria, which causes diseases of typhoid fever and intestinal infections and Shigella bacteria that cause diarrhea and post-operative wound infections [10,11]. It is necessary to pay attention to the heavy metal removal and microbial activity causing all these health problems and diseases using absorbent materials owing to their economic efficiency, environmental and low cost [12]. Nanostructures have high efficiency and speed in handling contaminated water compared to traditional materials, by a condition that, nanomaterials must meet many qualities to be non-toxic, natural substances with a high ability to remove heavy metals [13].

Chitosan is a natural polymer that has polymorphic properties as it is nontoxic, antibacterial, biocompatible and bio-degradable [14,15]. It is a linear polysaccharide composed of randomly distributed β—linked D-glucosamine and N-acetyl-D-glucosamine. Treating the chitin shells of shrimp and other crustaceans with an alkaline substance, like sodium hydroxide is an economic and beneficial substance for the environment for extracting Chitosan [16]. Being contain amine and hydroxide groups, Chitosan able to make strong bonds with ions [17]. Layered double hydroxides (LDHs) are an anionic or hydrotalcite-like clays, a class of lamellar compounds that comprise positive charged brucite as layers and hydrate exchangeable anions [18]. LDHs were remarkable because of their ability to adsorb heavy metals and use them as catalysts, catalyst supports, ion exchangers [19,20]. Intercalation process between LDHs and biomolecules creates bio-nano composite [21]. In this study, we focused on the synthesis of LDHs and the intercalation between Chitosan molecules. We used Chitosan-LDH Nano-composite as a bimodal, first for heavy metal removal–the main target-, second, as antimicrobial activity.

2. Materials and method

2.1. Chemicals and Strains

Shrimp peel were bought from market, $FeSO_4.7H_2O$, NaOH and Acetone were purchased from Piochem for chemicals (Egypt), HCl, $Al(NO_3)_2.9H_2O$, 1-Propanol, β- alanine, $CuCl_2.2H_2O$ were ordered from Oxford lab (India). $Zn(NO_3)_2.6H_2O$ was provided by SDFCL (India) while Methanol, Acetic Acid was purchased from Adwic (Egypt). All chemicals were used without purification and double-distilled water was used for preparation. Micro-organisms: Gram-positive organisms used in the study were *Bacillus subtillis*, *Lactobacillus sakei* LMG 2313, *Corynebacterium sp*, *Listeria inocua lmg* 2710 *Mycobacterium pheli*, *Enterococcus faecalis*, MRSA, *Staphylococcus aureus* 5247. Gram-negative organisms used were *Escherichia coli* ATCC 8739, *Escherichia coli* AO5, *Proteus Voulgaris* and *Pseudomonas auroginosa*. Yeast used was Candida albicans.

2.2. Preparation of chitosan

The method of preparation of the Chitosan is reported as found in the literature with some modifications [3,5,14]. Starting with Demineralization process for removing minerals, primarily calcium carbonate then Deproteinization process due to disruption of chemical bonds between chitin and proteins and using mechanical stirring in Decolouration process.

2.3. Synthesis of Al Zn-NO3 LDH–Chitosan

Chitosan-LDH nanocomposite was formed by composite by intercalation of Chitosan into the Al-Zn LDH by the co-precipitation method. 35ml of $Al(NO_3)_3.9H_2O$ (0.0004M) and $Zn(NO_3)_2.6H_2O$ were mixed with 25 ml distilled water, 10 ml 1-propanol and 30 ml of Chitosan (0.25g/l) using β-alanine (0.005 M) to increase Chitosan dissolution. NaOH solution (0.5 M) was added drop-wise with high magnetic stirring till pH (7.5); then, the solution was transferred into a Teflon-lined autoclave and heated at 120 for 18 hours. The precipitate was filtered, washed several times with distilled water to remove inorganic ions, and then dried in a vacuum at 60°for 24 hours [22–25].

2.4. Characterization of chitosan, LDH and chitosan-LDH

X-ray diffraction XRD patterns were determined on a 202964 PANalytical Empyrean X-ray diffraction using Cu Kα radiation scan angle 5°–80° range. Fourier transformation infra-red FTIR spectra were recorded using the KBr pellet technique on a Burker Vertex 70 FT-IR spectrometer in the range from 400 to 4000 cm^{-1}. The analysis of metals was measured by the atomic absorption spectroscopy AAS 200 series AA Agilent Technologies. High resolution transmission electron microscope HRTEM images were taken using (JEM-2100, JEOL, Japan) with an acceleration voltage of 200 kV. Field emission scanning electron microscope FESEM micrographs was obtained by Quanta FEG 250 (Switzerland).

2.5. Antimicrobial sensitivity test using agar diffusion method (Cup Technique)

Cup plate method is one of the official methods where the test samples diffuse from the cup through an agar layer in a Petri dish or plate until the growth of added microorganisms is restricted to a circular area or zone around the cavity containing an antibiotic solution [26]. The antimicrobial activity was expressed as zone diameter in millimeters [27].

2.6. Experimental procedure

Each overnight culture of the tested microorganisms was mixed with Muller Hinton agar media to give a final concentration of 1% microorganism (about 0.5 McFarland) and poured into sterile Petri dishes in a fixed amount of 20 ml using aseptic conditions. The sterile cork borer was used to prepare cups of 10 mm diameter. The tested samples and standard drugs with volumes of 60 µl were introduced into cups with the help of a micropipette. All the plates were kept at room temperature for effective diffusion of the test drug and standard and incubated at 37 ± 1°C for 24 hours. The inhibition zones around the cup indicated antibacterial activity. The diameter of the zone of inhibition was measured and recorded [28].

2.7. Determination of minimum inhibitory concentration (MIC)

Determination of minimum inhibitory concentration (MIC) was performed using agar dilution method according to Clinical Laboratory Standards Institute (CLSI) [29]. For each sample, different concentrations were diluted with Muller Hinton agar to give a final concentration ranging from (800 µg/ml–6.25 µg/ml). Dimethyl sulfoxide (DMSO) was used as a negative control plate. All isolated bacteria were subcultured on Brain Heart Infusion agar (B.H.I.A.) and incubated at 37 °C for 24 hours [30]. Three colonies of each microorganism were suspended in 5 ml saline, and the suspension was adjusted to a 0.5 McFarland standard and then diluted 10-fold with saline to give organism suspension of (1×106 to 5×106 CFU/ml). This suspension was then further diluted by putting 1 ml suspension to 9 ml saline to give a final suspension volume of 1×10^5 to 5×10^5 CFU/ml. A multiple inoculator was used to inoculate the prepared agar plates. A 100 µl (i.e. 10^4 CFU) of the prepared inoculums were put in the well of multi-inoculator, where each inoculation time by multi-inoculator gave about 10 µl of prepared inoculums to the plate (i.e. 103 CFU). Each experiment was performed in duplicates. All plates were incubated at 35°C for 48 hrs.

2.8. Metal adsorption study

Stock solutions (100 ppm) of copper (II) and iron (II) were diluted to give standard solutions of appropriate concentrations. Batch experiments were conducted in 100 ml conical flasks containing 50 ml of various concentrations of Cu^{2+} and Fe^{2+} at 300 rpm at room temperature. These experiments were conducted under different conditions. These conditions include: (i), the impact of pH (from 1 to 5) using 30 ppm of adsorbate and 40 mg of adsorbent for 20 hrs. (ii), the impact of adsorbent amount (from 10 to 50 mg) using pH 5 and 30 ppm of adsorbate for 20 hrs. (iii), the impact of time (from 30 to 360 minutes) using pH 5 and 40 mg of adsorbent and 30 ppm of adsorbate. After that, the adsorbent was separated by centrifugation and the supernatant was collected. The metal ion content was determined using Atomic Absorption Spectroscopy (AAS, 200 series AA Agilent Technologies). The adsorbed amount of metal ions (Q_e) onto the adsorbent was calculated by equation (1)

$$Q_e = \frac{c0 - ce}{M} V \tag{1}$$

Where C_0 and C_e (mg/l) are the initial and final concentration of the metal ion, V (Litre) is the volume of the aqueous solution and M (gm) is the weight of the adsorbent. The total ion removal efficiency was determined by equation (2).

$$(\%)\ Removal = \frac{C0 - Ce}{C0} \times 100 \tag{2}$$

3. Results and discussion

3.1. Materials characterization

The study comprises the synthesis of three materials Chitosan, LDH, Chitosan-LDH nanocomposite for waste water treatment and antimicrobial activity. FTIR spectra of all prepared materials are illustrated in Figure (1). Chitosan nanoparticles (Figure 1a), the band at 3455 cm^{-1} is related to O-H and NH stretching vibration [28]. The peaks at 1384 cm^{-1}, 1654 cm^{-1}, 11741 cm^{-1} and 1629 cm^{-1} can be attributed to the CH_3 symmetrical in an amide group, bending vibration of N-H group, carbonyl group and amide I, respectively [31]. As shown in Figure 1b, a strong broad absorption band appears at 3448 cm^{-1} which can be attributed to OH stretching [32]. The absorption band at 1382 cm^{-1} confirms the presence of NO_3^- group as an anion between the Zn-Al LDH layers [29]. Chitosan–LDH composite (Figure 1c), the absorption spectrum shows the presence of two bands at 3448, 1382cm^{-1} characteristic to both chitosan and LDH. A shift occurred in the peak at 1654 cm^{-1} due to the υ3 vibration of NO_3^-.

The crystal structure of both Chitosan and the Zn-Al LDH along with that of the Chitosan-LDH composite was studied by X-ray powder diffraction. As shown in Figure 2, the diffract gram of Chitosan nanoparticles is composed of two main peaks at approximately 9.4 ^{o} and 19.2^o which is similar to previously reported results [33]. The two peaks indicate a relative high degree of crystallinity for the prepared Chitosan nanoparticles [34]. The XRD pattern of Zn-Al LDH shows two intense peaks at 2Θ values of 11.7° and 23.2° corresponding to (003) and (006) plane, respectively (ICDD card 48-1022) [35]. The sharp peaks indicate high crystallinity of the prepared LDH nanoparticles. For the chitosan-LDH composite, it is assumed that the peaks for Chitosan has shielded those of LDH causing the noticed broadening in the (003) and (006) peaks. The peaks at 2Θ values of approximately 34.7 °, 39.2 °, 46.9 °, 60.4 ° and 61.8 ° correspond to the LDH plane (012), (015), (018), (110) and (113) respectively.

Figure 3a, shows FESEM micrograph of the Zn-Al-NO_3 LDH. Hexagonal like sheets of the LDH phase is clearly observed. In case of the Chitosan-LDH nano-composite (Figure 3b), the layered sheets of LDH are no longer hexagonal [18]. It can be also observed that the Chitosan nanoparticles are distributed over the surface of the LDH phase in the form of aggregates. Figure 4a shows a HRTEM micrograph of the Zn-Al-NO_3 LDH. It can be observed that the layers have a rectangle like shape with approximate dimensions of 430 x 400 nm. Figure 4b shows the homogenous distribution of Chitosan aggregates over the surface of the LDH phase in the Chitosan-LDH composite.

3.2. Metal adsorption study

The synthesized chitosan–LDHs composite was tested for removing Cu^{2+} and Fe^{2+} ions from aqueous solutions for water treatment.

3.2.1. Influence of pH on the efficiency of removal metal ions. The pH value is an important factor for the removal of heavy metals and adsorption of metal ions because it affects the degree of ionization of the solution, the surface charge of the adsorbate, the dissociation of the functional groups on the active sites of the adsorbent [36]. Moreover, the active sites can be protonated or deprotonated depending on the value of pH. Binding sites are deprotonated at high pH; thereby different functional groups are available for binding of metal ions producing attractive forces responsible for removal heavy metals from waste water [5,37]. At low pH, the sites of the adsorbent are positively charged where repulsion occurs between the metal ions and the adsorbent and consequently decreases the removal capacity. On the other hand, the removal of the metal depends on the direct relationship with the concentration of the metal. In the current study, Chitosan-LDH composite and an initial metal concentration are of 30 ppm. The optimum pH for both metals was found to be pH 5 with a maximum percentage removal of 54.06% and 91.5 for Cu^{2+} and Fe^{2+}, respectively. Moreover, increasing the pH above 6 leads to the precipitation of insoluble copper hydroxide and iron hydroxide, causing a decrease in the removal of Fe^{2+} and Cu^{2+} ions. Adsorption equilibrium studies were conducted under optimum conditions. As shown in figure 5, the removal efficiency of Cu^{2+} and Fe^{2+} ions increased with increasing pH solution where the removal efficiency of Fe^{2+} increased from 49% to 91.5% when the pH value was increased from 1 to 5 and the removal efficiency of Cu^{2+} increased from 6.6% to 54.06%.

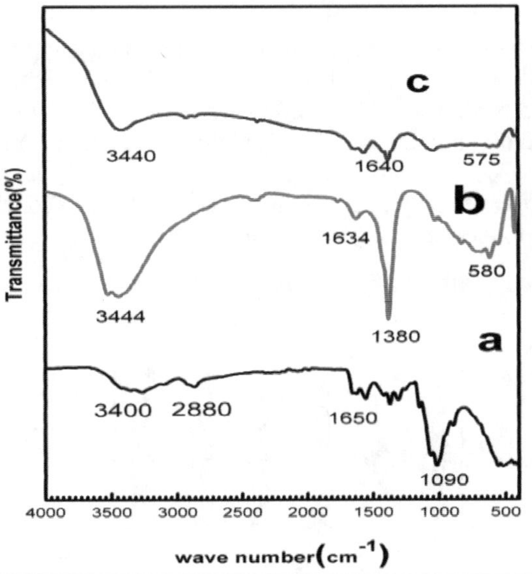

Figure 1. FTIR of (a) chitosan (b) LDH (c) chitosan-LDH.

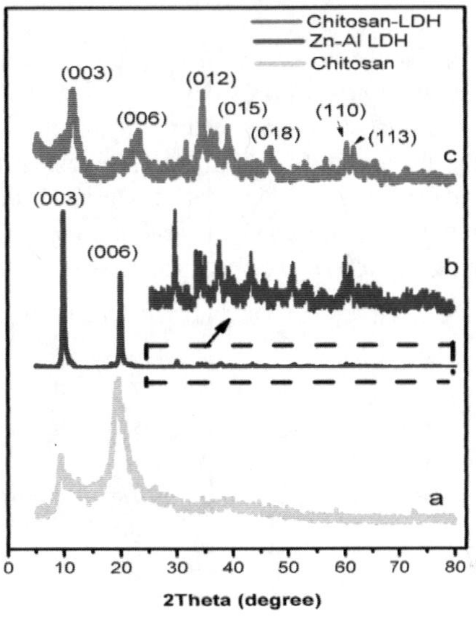

Figure 2. XRD patterns of (a) chitosan (b) LDH (c) chitosan-LDH.

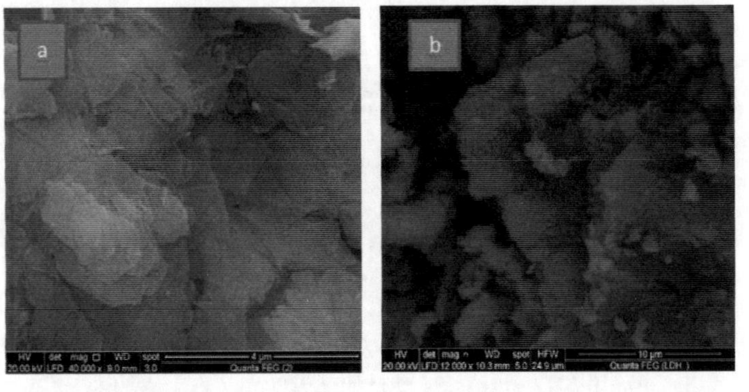

Figure 3. FESEM micrographs of (a) LDH (b) chitosan- LDH composite.

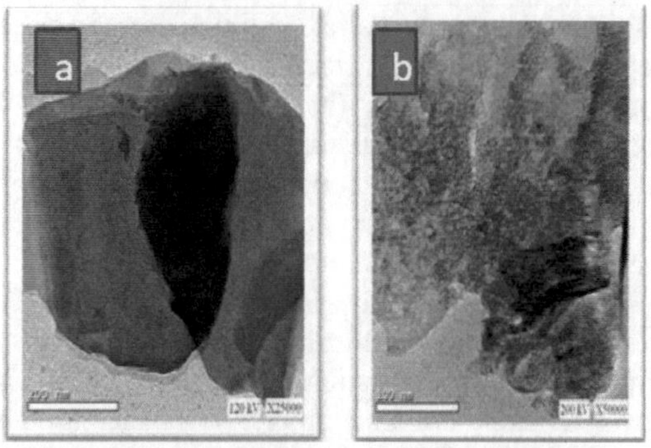

Figure 4. HRTEM micrographs of (a) LDH (b) chitosan-LDH composite.

3.2.2. Influence of contact time on the efficiency of removal metal ions. The effect of contact time was studied on removal efficiency of Cu^{2+} and Fe^{2+} ions by the Chitosan-LDH at pH 5, initial metal ion concentration of 30 ppm using 40 mg of adsorbent. The removal efficiency for both ions increased rapidly during the first 30 minutes reaching 32.6% and 55% for Cu^{2+} and Fe^{2+}, respectively and then onwards increased slightly as shown in Figure (6). The sharp increase in the first 30 minutes may be attributed to the presence of large number of vacant active sites on the surface of the nanocomposite. Increasing in removal of heavy metals is slight to the degree of equilibrium as shown in Figure 6.

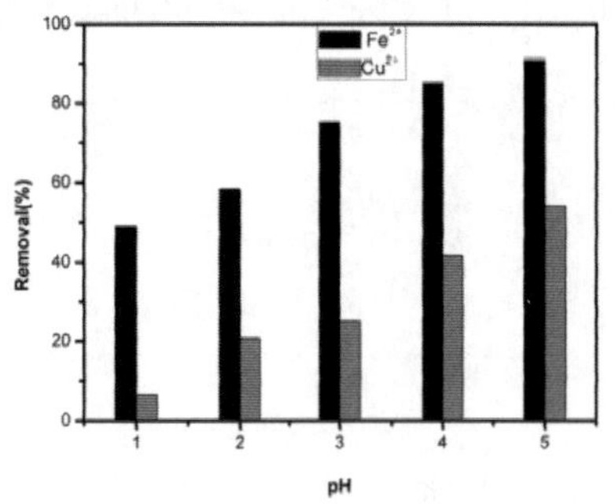

Figure 5. Influence of pH on the removal efficiency using (40 mg) and (30 ppm) at 20 h.

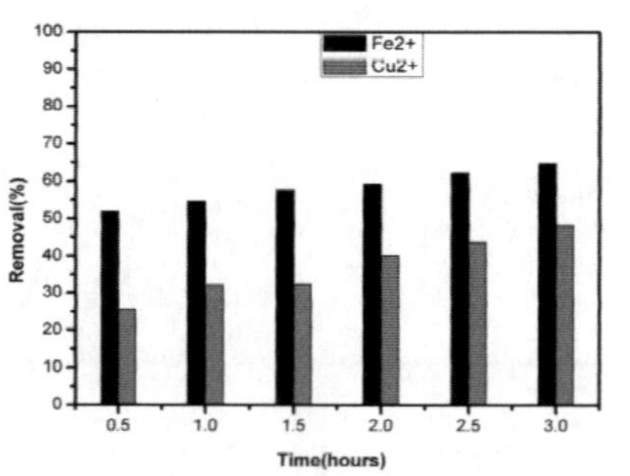

Figure 6. Influence of contact time on the removal efficiency using (40 mg) , (30 ppm) and (pH=5).

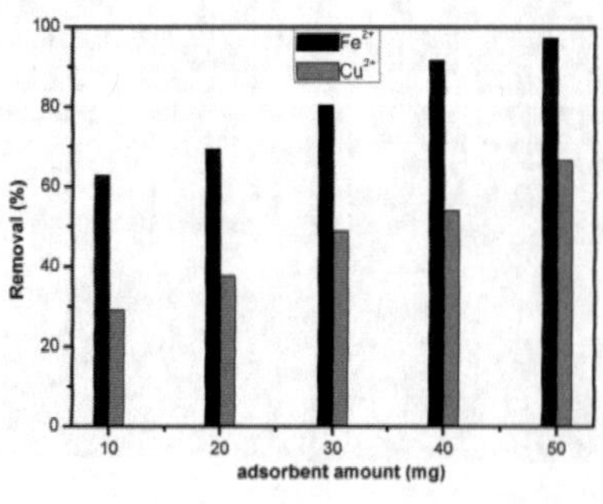

Figure 7. Influence of adsorbent amount on the removal efficiency using (30 ppm), (pH=5) at 20 h.

3.2.3. Influence of adsorbent amount on the efficiency of removal metal ions. The current study includes the effect of varying the adsorbent amount on the metal uptake for Fe^{2+} and Cu^{2+} from 0.01 to 0.05 gm, while keeping the other factors constant as pH = 5, concentration (30 ppm), constant time and at room temperature. The removal efficiency increased from 62.82 to 97.19% for Fe^{2+} and from 29.2 to 66.66% for Cu^{2+} ions when the adsorbent dose increased from 10 to 50 mg as shown in Figure 7; this increase can be attributed to the increase of the number of active sites on the surface area of the adsorbent with the increase of adsorbent dose.

3.2.4. Adsorption isotherm. Adsorption isotherms describe the relationship between the amount of the adsorbed substance per unit mass of adsorbent and its concentration in the equilibrium solution at constant temperature. They are used to describe adsorption equilibrium for waste water treatments. The equilibrium adsorption isotherms are essential to determine the adsorption capacity of Cu^{2+} and Fe^{2+} ions and to detect the nature of adsorption onto the Chitosan- LDH catalyst. Langmuir and Freundlich isotherms are the two most reliable models that can provide insights about the characteristics of the adsorbent surface. The Langmuir model assumes that the interaction occurs on the surface forming a mono-layer of adsorbate on the surface of adsorbent. The linear form of Langmuir equation is written as:

$$\frac{ce}{qe} = \frac{1}{q0\,kL} + \frac{1}{q0}ce \tag{3}$$

Which C_e (mg/l) is the equilibrium concentration of adsorbate, q_e (mg/g) the amount adsorbed, q_0 and k_L are Langmuir constants related capacity and rate of and adsorption, respectively. The values of C_e/q_e was plotted vs. C_e for both Cu^{2+} and Fe^{2+} as shown in Figure 8 and the model constants were estimated and listed in Table 1. Another important factor can be determined by Langmuir model is the separation factor or equilibrium Parameter, R_L can be expressed as:

$$R_L = \frac{1}{1+KLC0} \tag{4}$$

C_0 is the initial concentration of the metal ions (mg/g), the value of R_L shows to shape of the isotherm which can be favorable ($0<R<1$) or ($R=1$) linear, ($R=0$) irreversible, ($R>1$) unfavorable [38]. R_L values for Cu^{2+} and Fe^{2+} adsorption onto chitosan-LDH were calculated to be less than 1 and greater than zero, meaning that the adsorption was favorable. The Freundlich model assumes that the absorption occurs on a heterogeneous surface where the existing adsorption sites are not equivalent. The linear form of Freundlich equation is expressed as:

$$Ln\, qe = ln\, KF + \frac{1}{n} ln\ Ce \tag{5}$$

Where k_F is the adsorption capacity, N (n) is the isotherm exponent and (n) giving an indication of how favorable the adsorption process. 1/ n is the slope of the linear relation and its value ranging from 0 to 1. The surface becomes more heterogeneous as its value gets closer to 0. The Freundlich model parameters were estimated for both Cu^{2+} and Fe^{2+} through plotting ln q_e Vs ln C_e in Figure 9. However, the higher correlation coefficient of Cu^{2+} obtained from Langmuir model (R^2 = 0.9759) compared to Freundlich plot (R^2 = 0.9206) suggests the adsorption was taken place on the homogeneous surface of the adsorbent much better than the multilayer adsorption. While, for Fe^{2+}, the higher correlation coefficient of Fe^{2+} was from Langmuir model (R^2=0.7495) compared to Freundlich plot (R^2 =0.8441) suggesting the adsorption was occurred on the heterogeneous surface of the adsorbent much better than the monolayer adsorption.

As shown in Table 2, the value of R^2 for Cu^{2+} adsorption over the surface of the Chitosan-LDH composite was high for Langmuir model compared to that of Freundlich model showing a better fit of the data. The opposite is true for the case of Fe^{2+} showing the possibility of multilayer adsorption over the surface of the Chitosan-LDH composite.

3.2.5. Adsorption kinetics analysis. Studying the adsorption kinetics, describes the removal rate of the solute, and this rate controls the residence time of removal the adsorbate at the solid solution interface including the diffusion process. The mechanism of adsorption depends on the physical and chemical characteristics of the adsorbent. The adsorption kinetics explains the solute uptake on the adsorbent and its depended on the contact time that ranged from 30 to 180 minutes for removal of Cu^{2+} and Fe^{2+}. Figure 10 displays the kinetics of metal ions adsorption using pseudo first order and pseudo second order. The pseudo-first kinetic order assumes the physical sorption binding according the following equation:

$$Ln(qe - qt) = ln\, qe - K1t \tag{6}$$

Where, q_e and q_t are the amounts of metal on the adsorbent at equilibrium and at the time. K_1 is the pseudo-first adsorption rate constant. q_e and k_1 are the slope and intercept on the linear plot by the relation between time (min) and ln (q_e - q_t). While, the pseudo second order equation is written as:

$$\frac{t}{qt} = \frac{1}{K2\ qe^2} + \frac{t}{qe} \tag{7}$$

Where, K_2 is pseudo second adsorption rate constant (min^{1-}) and. K_2 is calculated from the slope of the linear plot t/q_t versus t (min).

Table 1. Langmuir and Freundlich isotherm factors for Cu^{2+} and Fe^{2+} .

Isotherm	Cu^{2+}	Fe^{2+}
Langmuir model :		
q_0 (mg/g)	23.0733	5.0403
k_L(L/mg)	0.2672	0.1902
R_L	0.0696	0.0951
R^2	0.9759	0.7495
Freundlich model:		
K_f	9.6234	25.0056
1/n	0.2249	0.2726
N	4.4464	3.6683
R^2	0.9206	0.8441

Table 2. Kinetic factors of pseudo first order and pseudo second order for removing Cu^{2+} and Fe^{2+}.

Order model	Cu^{2+}	Fe^{2+}
Pseudo first order:		
q_e calc(mg/g)	18.137	16.0225
q_e exp(mg/g)	16.875	34.35
k_1(min^{-1})	-0.021	-0.125
R^2	0.9024	0.9914
Pseudo second order:		
q_e calc(mg/g)	21.231	25.57
q_e exp(mg/g)	16.875	34.35
k_2(min^{-1})	$1.28*10^{3-}$	$2.85*10^{3-}$
R^2	0.9958	0.9523

IOP Conf. Series: Materials Science and Engineering **464** (2018) 012005 doi:10.1088/1757-899X/464/1/012005

According to Table (2), q_e, k_1, K_2 and R_2 (the correlation coefficient) from the linear plots, For Cu^{2+}, the pseudo-second-order linear plots revealed higher R^2 values than the pseudo-first-order while For Fe^{2+}, the pseudo-first order linear plots higher R^2 values than the pseudo-second order. The results showed that the adsorption of Cu^{2+} followed pseudo second order and controlled by chemical adsorption while Fe^{2+} adsorption was physical and followed first order.

3.3 Antimicrobial activity

The antibacterial activity was studied for Chitosan, LDH, and Chitosan-LDH composite against Gram-positive and Gram–negative by agar disc diffusion method. Table (4) displays the measurements of inhibition zone. The result indicated that Chitosan-LDH composite exhibited more antimicrobial activity compared to Chitosan and LDH as parents. The inhibition zone against Lactobacillus sakei LMG was (15mm), Corynebacterium sp (16mm), Staphylococcus aureus (16mm), Escherichia coli AO5 (14mm). The high efficiency of Chitosan-LDH composite was due to the large surface area that provides better contact with microorganisms. Table 5 exhibits the (MIC) values of Chitosan, LDH, and Chitosan-LDH composite, the result showed that the lowest concentration of the tested materials inhibited the growth of Gram–positive bacteria i.e. the tested materials possessed strong antibacterial activity against Gram-positive bacteria The (MIC) value of Chitosan- LDH against Corynebacterium sp was less than 6.25 µg/mL.

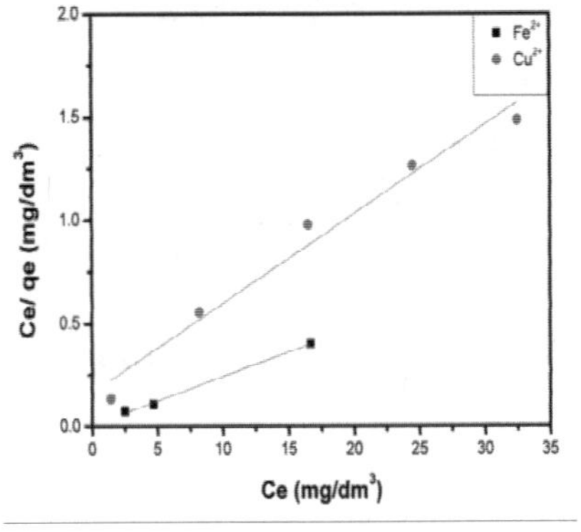

Figure 8. Langmuir adsorption isotherm graphs for removing Cu^{2+} and Fe^{2+}.

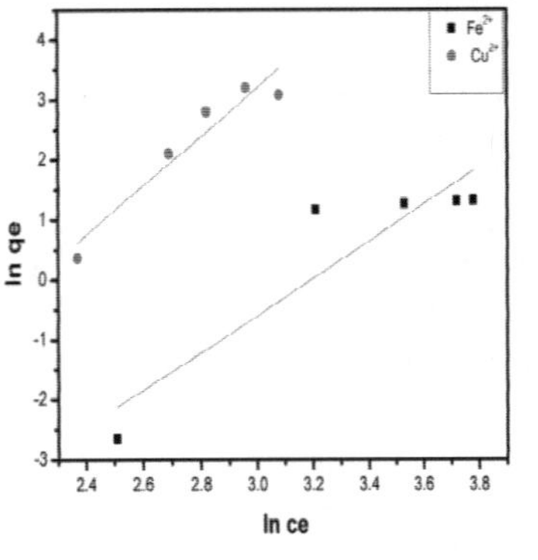

Figure 9. Freundlich adsorption isotherm graphs for removing Cu^{2+} and Fe^{2+}.

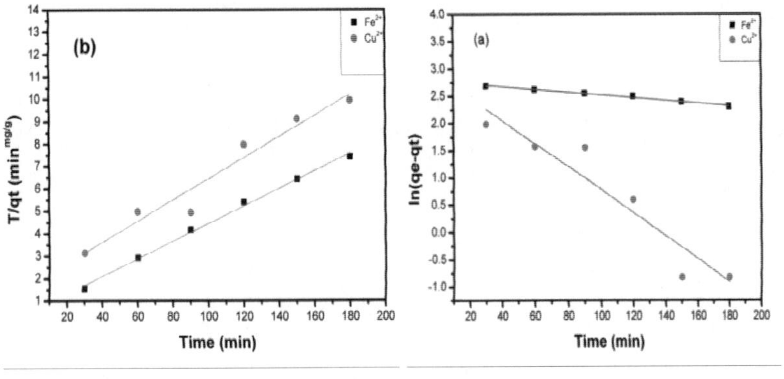

Figure 10. Kinetic plot (a) pseudo first order (b) pseudo second order for Cu^{2+} and Fe^{2+}.

4. Conclusion

In this study, Chitosan-LDH composite was chosen as an adsorbent for removing metal ions from waste water. It approved satisfactory results for Cu^{2+} and Fe^{2+}. The removal efficiency of metal ions was affected by the pH; the adsorption of Cu^{2+} and Fe^{2+} ions increased with the increase of the pH values from 1 to 5. The pH 5was the optimum pH for maximum adsorption for iron and copper, initial concentration was 30 ppm and 40 mg adsorbent. Kinetic studies suggested that the adsorption of Cu^{2+} by Chitosan-LDH composite was chemical adsorption and followed the second-order kinetics model while Fe^{2+} was physical and followed first order kinetics model. The removal of iron was better than copper; the removal of iron ions achieved to 99.3%, but for copper achieved to 85.6%. Chitosan–LDH composite was tested for its efficiency against some species of microorganisms such as different gram positive (Lactobacillus sakei LMG 2313, Corynebacterium sp, Staphylococcus aureus 5247) and gram negative bacteria (Escherichia coli AO5). Strong antibacterial activity against Gram-positive bacteria achieved; where the lowest concentration of antimicrobial/antifungal agent inhibited the growth of Gram positive bacteria or giving no visible growth.

Table 3. Comparing our results with previous studies on the same metal ions.

Adsorbent	Metal ion	% Removal	pH	Reference
polyaniline graft chitosan beads	Cu^{2+}	79.8	5	[2]
Nano chitosan / Soduim ALGINATE/Microcrystalline cellulose	Cu^{2+}	76	6	[31]
Chitosan-LDH	Cu^{2+}	85.6	5	Current study
Chitosan-LDH	Fe^{2+}	99.3	5	Current study
Synthesized Chitosan	Fe^{2+}	72	8	[39]

Table 4: The zone of inhibition was measured by a scale and the measurements in mm are tabulated.

	Gram positive rods					Gram positive cocci			Gram negative rods				Yeast
	Bacillus subtilis	*Lactobacillus sakei LMG 2313*	*Corynebacterium sp.*	*Listeria inocua lmg 2710*	*Mycobacterium pheli*	*Enterococcus faecalis*	*MRSA*	*Staphylococcus aureus 5247*	*Escherichia coli ATCC 8739*	*Escherichia coli AO5*	*Psedomonas aeruginosa*	*Proteus voulgaris*	*Candida albicans*
Chitosan	--	--	--	--	--	--	--	--	--	--	--	--	--
LDH	--	--	--	--	--	--	--	--	--	--	--	--	--
LDH-Chitosan	--	15	16	--	--	--	--	16	--	14	--	--	--

Table 5. the MIC was measured as μ g/ml.

	Gram positive rods	Gram positive cocci	Gram negative rods	Yeast

	Bacillus subtilis	*Lactobacillus sakei LMG 2313*	*Corynebacterium sp.*	*Listeria inocua lmg 2710*	*Mycobacterium pheli*	*Enterococcus faecalis*	*MRSA*	*Staphylococcus aureus 5247*	*Escherichia coli ATCC 8739*	*Escherichia coli AO5*	*Psedomonas aeruginosa*	*Proteus voulgaris*	*Candida albicans*
Chitosan	>800	800	400	>800	>800	>800	>800	>800	>800	>800	>800	>800	>800
LDH	>800	800	400	>800	>800	>800	>800	>800	>800	>800	>800	>800	>800
LDH-Chitosan	>800	12.5	< 6.25	>800	>800	>800	>800	400	>800	200	>800	>800	>800

Also, Chitosan-LDH composite exhibited good antimicrobial potency with MIC value < than 6.25 μg/mL. In a nutshell, Chitosan-LDH composite has evidenced that it is a good adsorbent for removal heavy metals and a good antimicrobial for water treatment process.

References

[1] Alimohammadi V, Sedighi M and Jabbari E 2017 Experimental study on efficient removal of total iron from wastewater using magnetic-modified multi-walled carbon nanotubes *Ecol. Eng.* **102** 90–7

[2] Igberase E, Osifo P and Ofomaja A 2014 The adsorption of copper (II) ions by polyaniline graft chitosan beads from aqueous solution: equilibrium, kinetic and desorption studies *J. Environ. Chem. Eng.* **2** 362–9

[3] Wan M-W, Kan C-C, Rogel B D and Dalida M L P 2010 Adsorption of copper (II) and lead (II) ions from aqueous solution on chitosan-coated sand *Carbohydr. Polym.* **80** 891–9

[4] Ngah W S W, Ab Ghani S and Kamari A 2005 Adsorption behaviour of Fe (II) and Fe (III) ions in aqueous solution on chitosan and cross-linked chitosan beads *Bioresour. Technol.* **96** 443–50

[5] Sivakami M S, Gomathi T, Venkatesan J, Jeong H-S, Kim S-K and Sudha P N 2013 Preparation and characterization of nano chitosan for treatment wastewaters *Int. J. Biol. Macromol.* **57** 204–12

[6] Al-Karawi A J M, Al-Qaisi Z H J, Abdullah H I, Al-Mokaram A M A and Al-Heetimi D T A 2011 Synthesis, characterization of acrylamide grafted chitosan and its use in removal of copper (II) ions from water *Carbohydr. Polym.* **83** 495–500

[7] Ngah W S W, Teong L C and Hanafiah M 2011 Adsorption of dyes and heavy metal ions by chitosan composites: A review *Carbohydr. Polym.* **83** 1446–56

[8] Esmat M, Farghali A A, Khedr M H and El-Sherbiny I M 2017 Alginate-based nanocomposites for efficient removal of heavy metal ions *Int. J. Biol. Macromol.* **102** 272–83

[9] Ngah W S W and Fatinathan S 2008 Adsorption of Cu(II) ions in aqueous solution using chitosan beads, chitosan–GLA beads and chitosan–alginate beads *Chem. Eng. J.* **143** 62–72

[10] Jiang Y, Gong J, Zeng G, Ou X, Chang Y, Deng C-H, Zhang J, Liu H-Y and Huang S-Y 2016 Magnetic chitosan – graphene oxide composite for anti-microbial and dye removal applications *Int. J. Biol. Macromol.* **82** 702–10

[11] Banerjee S L, Khamrai M, Sarkar K, Singha N K and Kundu P P 2016 Modified chitosan encapsulated core-shell Ag Nps for superior antimicrobial and anticancer activity *Int. J. Biol. Macromol.* **85** 157–67

[12] He X, Qiu X and Chen J 2017 Preparation of Fe(II)–Al layered double hydroxides: Application to the adsorption/reduction of chromium *Colloids Surfaces A Physicochem. Eng. Asp.* **516** 362–74

[13] Wang X, Guo Y, Yang L, Han M, Zhao J and Cheng X 2012 Nanomaterials as sorbents to

remove heavy metal ions in wastewater treatment *J. Environ. Anal. Toxicol* **2** 1000154
[14] Negm N A, El Sheikh R, El-Farargy A F, Hefni H H H and Bekhit M 2015 Treatment of industrial wastewater containing copper and cobalt ions using modified chitosan *J. Ind. Eng. Chem.* **21** 526–34
[15] Muhammed R, Junise V, Saraswathi P, Krishnan P and Dilip C 2010 Development and characterization of chitosan nanoparticles loaded with isoniazid for the treatment of tuberculosis *Res J Pharm Biol Chem Sci* **1** 383–90
[16] Ribeiro L N M, Alcântara A C S, Darder M, Aranda P, Araújo-Moreira F M and Ruiz-Hitzky E 2014 Pectin-coated chitosan-LDH bionanocomposite beads as potential systems for colon-targeted drug delivery *Int. J. Pharm.* **463** 1–9
[17] Nejati K, Keypour H, Nezhad P D K, Rezvani Z and Asadpour-Zeynali K 2015 Preparation and characterization of cetirizine intercalated layered double hydroxide and chitosan nanocomposites *J. Taiwan Inst. Chem. Eng.* **53** 168–75
[18] Abdallah S H 2014 Kinetic Controlled Release Study of Salicylate Anion from Synthesized and Characterized of Ni/Al-salicylate-Layered Double Hydroxide Nanohybrid *Journals kufa chamical* 26–33
[19] Chang Z, Wu C, Song S, Kuang Y, Lei X, Wang L and Sun X 2013 Synthesis mechanism study of layered double hydroxides based on nanoseparation *Inorg. Chem.* **52** 8694–8
[20] Zhao Y, Xiao F and Jiao Q 2011 Hydrothermal synthesis of Ni/Al layered double hydroxide nanorods *J. Nanotechnol.* **2011**
[21] Moaty S A A, Farghali A A, Moussa M and Khaled R 2017 Remediation of waste water by Co–Fe layered double hydroxide and its catalytic activity *J. Taiwan Inst. Chem. Eng.* **71** 441–53
[22] Mahjoubi F Z, Khalidi A, Abdennouri M and Barka N 2017 Zn–Al layered double hydroxides intercalated with carbonate, nitrate, chloride and sulphate ions: Synthesis, characterisation and dye removal properties *J. Taibah Univ. Sci.* **11** 90–100
[23] Moaty S A A, Farghali A A and Khaled R 2016 Preparation, characterization and antimicrobial applications of Zn-Fe LDH against MRSA *Mater. Sci. Eng. C* **68** 184–93
[24] Elgiddawy N, Essam T M, El Rouby W M A, Raslan M and Farghali A A 2017 New approach for enhancing Chlorella vulgaris biomass recovery using ZnAl-layered double hydroxide nanosheets *J. Appl. Phycol.* 1–9
[25] Mahmoud R, Moaty S A, Mohamed F and Farghali A 2017 Comparative Study of Single and Multiple Pollutants System Using Ti–Fe Chitosan LDH Adsorbent with High Performance in Wastewater Treatment *J. Chem. Eng. Data* **62** 3703–22
[26] Seeley H W, VanDemark P J and Lee J J 1975 Microbes in action : a laboratory manual of microbiology *Microbes action. A Lab. Man. Microbiol.* 450
[27] Ahmed S H, Amin M A, Saafan A E, El-Gendy A O and ul Islam M 2017 Measuring susceptibility of Candida albicans biofilms towards antifungal agents *J. Microbiol. Biotechnol. Res.* **3** 149–56
[28] El-Gendy A O, Essam T M, Amin M A, Ahmed S H and Nes I F 2013 Clinical screening for bacteriocinogenic Enterococcus faecalis isolated from intensive care unit inpatient in Egypt *J. Microb. Biochem. Technol* **4** 161–7
[29] Sader H S, Flamm R K and Jones R N 2013 Antimicrobial activity of daptomycin tested against Gram-positive pathogens collected in Europe, Latin America, and selected countries in the Asia-Pacific Region (2011) *Diagn. Microbiol. Infect. Dis.* **75** 417–22
[30] Tarek N, Hassan H M, AbdelGhani S M M, Radwan I A, Hammouda O and El-Gendy A O 2014 Comparative chemical and antimicrobial study of nine essential oils obtained from medicinal plants growing in Egypt *Beni-Suef Univ. J. Basic Appl. Sci.* **3** 149–56
[31] Vijayalakshmi K, Gomathi T, Latha S, Hajeeth T and Sudha P N 2016 Removal of copper (II) from aqueous solution using nanochitosan/sodium alginate/microcrystalline cellulose beads *Int. J. Biol. Macromol.* **82** 440–52
[32] Ahmed A A A, Talib Z A, Bin Hussein M Z and Zakaria A 2012 Zn-Al layered double hydroxide prepared at different molar ratios: Preparation, characterization, optical and

dielectric properties *J. Solid State Chem.* **191** 271–8

[33] Wang Y, Pitto-Barry A, Habtemariam A, Romero-Canelon I, Sadler P J and Barry N P E 2016 Nanoparticles of chitosan conjugated to organo- ruthenium complexes *Inorg. Chem. Front.* **3** 1058–64

[34] Qi L, Xu Z, Jiang X, Hu C and Zou X 2004 Preparation and antibacterial activity of chitosan nanoparticles *Carbohydr. Res.* **339** 2693–700

[35] Qu J, He X, Li X, Ai Z, Li Y, Zhang Q and Liu X 2017 Precursor preparation of Zn-Al layered double hydroxide by ball milling for enhancing adsorption and photocatalytic decoloration of methyl orange *RSC Adv.* **7** 31466–74

[36] Yang F, Sun S, Chen X, Chang Y, Zha F and Lei Z 2016 Mg–Al layered double hydroxides modified clay adsorbents for efficient removal of Pb 2+ , Cu 2+ and Ni 2+ from water *Appl. Clay Sci.* **123** 134–40

[37] Lin Y, Hong Y, Song Q, Zhang Z, Gao J and Tao T 2017 Highly efficient removal of copper ions from water using poly (acrylic acid)-grafted chitosan adsorbent *Colloid Polym. Sci.* **295** 627–35

[38] Martins A C, Pezoti O, Cazetta A L, Bedin K C, Yamazaki D A S, Bandoch G F G, Asefa T, Visentainer J V. and Almeida V C 2015 Removal of tetracycline by NaOH-activated carbon produced from macadamia nut shells: Kinetic and equilibrium studies *Chem. Eng. J.* **260** 291–9

[39] Hadi A G 2016 Removal of Fe (II) and Zn (II) ions from aqueous solutions by synthesized chitosan *Int. J. ChemTech Res.* **9** 343–9

Characterization and electrochemical performance of Manganese oxide with carbon nano tubes as negative electrode for lithium batteries

Atef YShenouda

Central Metallurgical Research and Development Institute (CMRDI), Tebbin, P.O. Box 87 Helwan, Egypt. E-mail: ayshenouda@gmail.com

Abstract. Different ratios of Mn_3O_4/CNT (carbon nano tubes) samples were prepared by ultrasonic and hydrothermal reactions. The added ratios of CNT: Mn wt./wt. were 0, 5, 15, 25, 35 and 50%. The phases of the samples were identified by X-ray diffraction (XRD). The particle morphology was studied with a field emission scanning electron microscope (FESEM). The lowest charge transfer resistance, R_{ct} was 19.18 Ω with cell of Mn_3O_4/35% CNT. For Li/ Mn_3O_4- 35% CNT cell, the specific discharge capacity was about **688** $mAhg^{-1}$ after **100** cycles, which was the highest value among the other cells at discharge current density $0.02Acm^{-2}$.

Keywords: Lithium battery anode, Mn_3O_4/CNT.

1. Introduction

Small and Portable electronic devices such as laptops, mobile telephones, camcorders, etc. use Lithium-ion battery (LIB). They are considered the essential electric power component for these devices due to their excellent properties like long durability, high energy density, easily handling and safe for the environment. A variety of metals and metal oxides are intensively studied in order to obtain high-capacity anode materials for LIBs [1–4].

The synthesis of Mn_3O_4 with CNT was reported in some researches [5–16]. Manganese oxide with multi-wall carbon nanotubes (MWCNTs) was prepared by hydrothermal method using CH_3COONa. Mn3O4 nano-crystals were introduced inside MWCNTs [6]. The nanocomposite performed good cycling stability and rate capability. The initial reversible capacity was 1380 mAh g−1 at a current density of 100 mAg^{-1}. The reversible cycle discharge capacity was 592 $mAhg^{-1}$ after 50 cycles. The good performance of Mn_3O_4/MWCNTs was discussed.

It was reported that Mn_3O_4nano-particles anchored on continuous super-aligned carbon nanotube (SACNT) films were obtained by decomposition of $Mn(NO_3)_2$[9]. The electrochemical performance of the Mn_3O_4/SACNT composite electrode is greatly affected by the grain size of Mn_3O_4. For smaller particle size, the electrode displayed smaller voltage hysteresis and much improved reversibility and rate capability. Mn_3O_4 with particle size of less than 10 nm inside the Mn_3O_4/SACNT electrodes showed a capacity of 342 mAh g^{-1} at 10 C. This composite material has more advantages than the commercial graphite anode and promising in practical applications.

Mn_3O_4 / CNT was synthesized by a simple hydrothermal method [10]. MnO / CNT was obtained by heating Mn_3O_4/CNTs at 500°C for 3 h in flowing Ar/H_2. After 30 cycles, the discharge capacity of Mn3O4 electrode is only 280 $mAhg^{-1}$, whereas that of Mn_3O_4/CNTs electrode can still remain with 548 mAh g-1 at current density 100 mAh g^{-1}. While MnO / CNT cell has 600 vs. 200 mAh g^{-1} for MnO. The manganese oxide/CNTs composites has stable cyclability, good rate performance and promising anode materials.

Carbon nanofiber/Mn3O4 (CNF/Mn_3O_4) coaxial nano-fibers with a three-dimensional (3D) structure were prepared for lithium ion batteries by electrophoretic deposition on an electrospun CNF cathode followed by heat treatment in air [11]. The CNF/Mn_3O_4 coaxial nanocables showed a high reversible capacity of 760 mAh g^{-1} at 100 mA g^{-1} even after 50 charge discharge cycles without capacity fading.

A facile and green chemistry route was developed to prepare a Mn_3O_4 nanoparticles (NPs)/ exfoliated graphite (EG) composite by artfully tuning the traditional fabrication process of EG [12]. During this treatment, Mn_3O_4-NPs with high crystallinity and uniform dimension of 7 nm were found to be homogeneously and firmly anchored on the surface of EG. The composite as an anode material of Li-ion batteries exhibited suitable specific discharge capacity of 655 $mAhg^{-1}$ extending to more than 100 cycles.The aim of this work is to study the effect of different ratios of CNT on the performance of Mn – oxide in lithium cell as there is a need to study effect of the CNT ratios according to the reported literature.

2. Experimental

2.1. Materials preparation and characterization

Different ratios of carbon nano tubes (CNT), (99%, Ad-Nano Technologies Private Limited, India) 0, 5, 15, 25, 35 and 50% were added to stoichiometric amounts of $(CH_3CO_2)_2Mn.4H_2O$ (Aldrich, 99.5%) dissolved in distilled water to prepare the samples. The ratios of CNT were calculated as Mn: C. The suspended CNT of each mixture solution was stirred and heat-treated at 80°C for 2 hours. Furthermore, the mixture was exposed toultrasonic vibration for one hour. After that, the mixture was transferred to an autoclave vessel (250 cm^3) and heat-treated at 200°C for 24 hours.

The crystalline phases of the samples were identified by X-ray diffraction (XRD) using a step size of 0.02° and at a scanning rate of 0.16°/min on a Brucker axis D8 diffractometer with crystallographic data software Topas 2 using Cu Kα (λ = 1.54056 Å) radiation operating at accelerating voltage and applied current were 40 kV and 80 mA, respectively. The diffraction data was recorded for 2θ values between 10° and 80°. The particle morphology was studied with a field emission scanning electron microscope (FESEM QUANTAFEG 250). The magnetic properties of some of the samples powders were measured at room temperature using a vibrating sample magnetometer (VSM, 7410 Lakeshore, USA) in a maximum applied field of 20 kOe.

2.2. Electrochemical measurements

The homogeneous slurry used to form the electrodes was composed of 85 wt% active materials, 10 wt% acetylene black, and 5 wt% sodium carboxy methyl cellulose (SCMC) as binder dissolved in distilled water. It was then spread onto Cu foil substrates. The area of each coated electrode was 1 cm^2. The electrodes were dried in a vacuum oven at 110°C for 12 h. The electrodes were then pressed at a pressure of 2000 kg/cm2. The active material loading was about 5 mg for each individual electrode. CR2032 coin cells were then assembled in an argon filled glove box (Mbraun, Unilab, Germany), using lithium metal foil as the counter electrode. The electrolyte was 1M $LiPF_6$ in a mixture of ethylene carbonate (EC) and dimethyl carbonate (DMC) (1:1 by volume, provided by MERCK). The cells were galvanostatically charged and discharged over a voltage range of 0.0–3 V using a current of 0.02 A/ cm^2 for both processes. Cyclic voltammetry (CV) measurements were performed using a Multi-stat CHI660 Electrochemical Workstation at a 0.1 mVs^{-1} scanning rate, and the potential windows were 0 and 3 V versus Li/Li+ electrode. The Galvanostatic charge-discharge measurements were performed on a Neware Battery Tester. The frequency range of the AC impedance measurement was 1MHz – 10 mHz and10 mV amplitude.

3. Results and Discussion

3.1 Structural Characterization

X-ray diffraction patterns of the manganese oxide and manganese oxide with CNT samples show suitable crystallinity as shown in Figure 1. Samples diffraction peaks exhibited good crystalline structures. Their structures were indexed to the tetragonal system with space group of I41/amd (JCPDS No. 24–0734). The peak at 2θ =26.4° can be achieved to (002) of the hexagonal carbon in CNT (JCPDS No. 26–1079). Mainly, there is no other phase that observed in the patterns. The main diffractions are (103), (211) and (224) for Mn3O4. The crystallite sizes of Mn_3O_4, Mn_3O_4 with 25% CNT and Mn3O4 with 35% CNT are 22.07, 13.58 and 11.03 nm, respectively at 2 θ =37° according to the Debye-Scherrer equation [2,3]:

$$L = 0.94\,\lambda / w \cos\theta \qquad (1)$$

where θ and w are the Bragg angle and the full width at half maximum peak, measured in radians, of each diffraction peak, respectively. Also, λ is the X-ray wavelength (1.54056Å), and L is the effective particle or grain size.

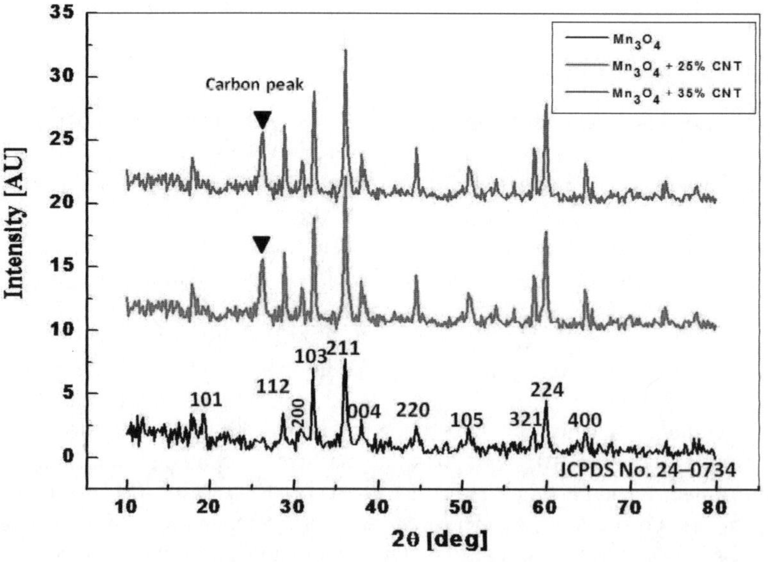

Figure 1. XRD of (a) Mn_3O_4, (b) Mn_3O_4 + 25% CNT and (c) Mn_3O_4 + 35% CNT.

Figure 2 shows a FESEM image of the samples. The powder of Mn_3O_4 itself has an average particle size of ~1 µm as shown in Figure 2(a). On the other hand, the small nano particles of Mn_3O_4 entered inside CNT particles and formed morphology like distal colon of rabbit due to sonication and hydrothermal treatment that degraded the particles of Mn3O4. So, CNT diameter without Mn_3O_4 has a diameter of 41.2 nm (Figure 2b) but samples with 25 and 35% CNT have 108.4 and 88.6 nm, respectively as shown in Figure 2c,d. Thus, 35% CNT has less particle size than 25% CNT. The preparation of Mn_3O_4 without using CNT gives agglomeration of the grains with large particle size. However, the use of CNT enhances the formation of nano-particles of Mn_3O_4 under ultrasonic technique. Similar reported results were published explaining that HRTEM (high resolution transmission electron microscope) images indicated the presence of the Mn_3O_4nano-particles enclosed and encapsulated within the carbon nanotubes [4].

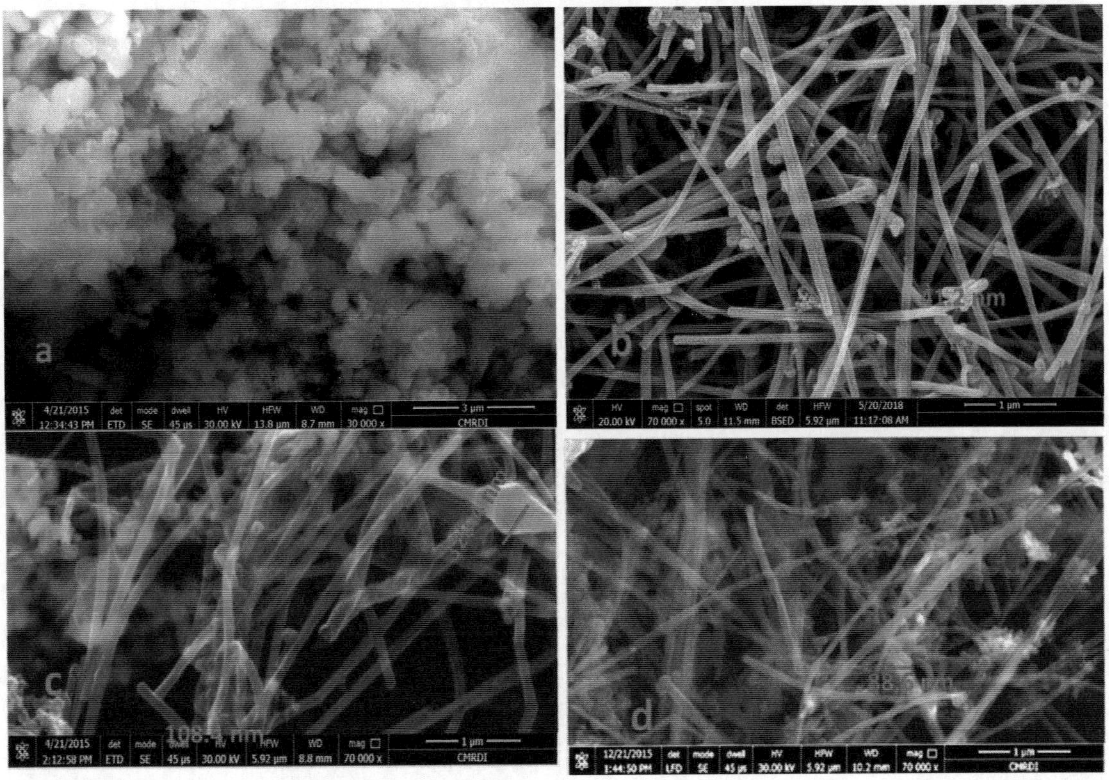

Figure 2. SEM of (a) Mn_3O_4, (b) CNT, (c) Mn_3O_4 + 25% CNT & (d) Mn_3O_4 + 35% CNT.

3.2. Electrochemical Impedance Spectroscopy and magnetic properties measurements

Electrochemical impedance spectroscopy (EIS) may be considered as one of the most sensitive tools for the study of differences in electrode behavior due to surface modification. The electrochemical impedance spectra of the cells, as presented in Figure 3, show an intercept at high frequency on the real axis Z′ for the resistance of the electrolyte, R_e, followed by a semicircle in the high-middle frequency region, and a straight line in the low frequency region. The numerical value of the diameter of the semicircle on the Z_{real} axis is approximately equal to the charge transfer resistance, R_{ct}, and therefore, it can be seen that there is a marked decrease in Rct with the addition of CNT until 35% (R_{ct} =127 Ω), but 50% CNT increases R_{ct}, 928 Ω. So, the increase of CNT percentage decreases the R_{ct} until 35% CNT and after that the charge resistance increases again with the increase of CNT. i.e. 35% CNT is the optimum condition of these experiments.

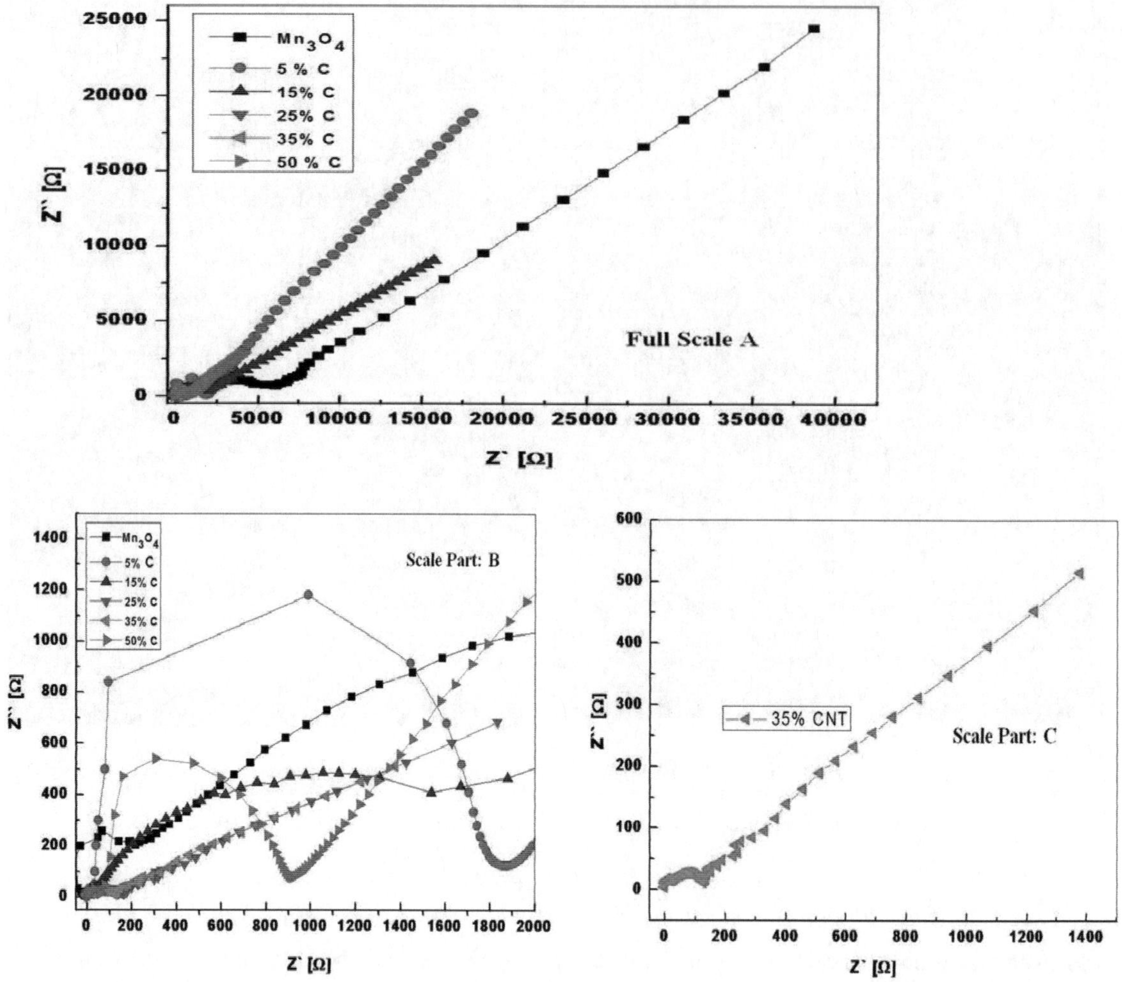

Figure 3. EIS of cells prepared from: (a) Mn_3O_4, (b) Mn_3O_4 + 5% CNT, (c) Mn_3O_4 + 15% CNT, (d) Mn_3O_4 + 25% CNT, (e) Mn_3O_4 + 35% CNT and (f) Mn_3O_4 + 50% CNT. Full Scale: A Scale Part B up to Z` = 2000 Ω and Scale Part C for Mn_3O_4 + 35% CNT only.

The straight line in the low frequency region is attributed to the diffusion of the lithium ions into the bulk of the electrode material, or the so-called Warburg diffusion. The plot of the real part of the impedance, Z re, versus the reciprocal root square of the lower angular frequencies is displayed in Figure 4.

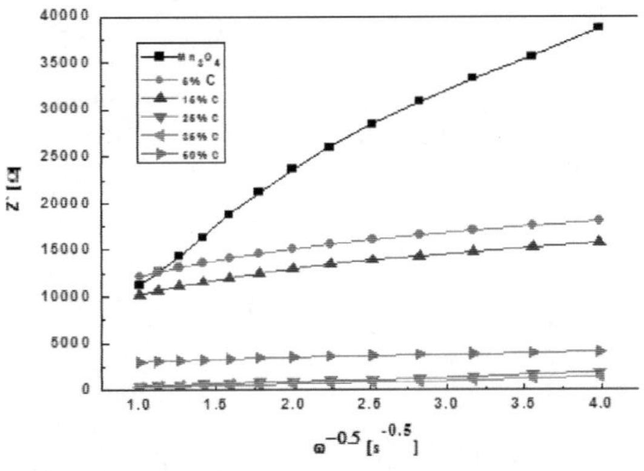

Figure 4. Relationship between real impedance with the low angular frequencies for Li/ Mn_3O_4-CNT cells.

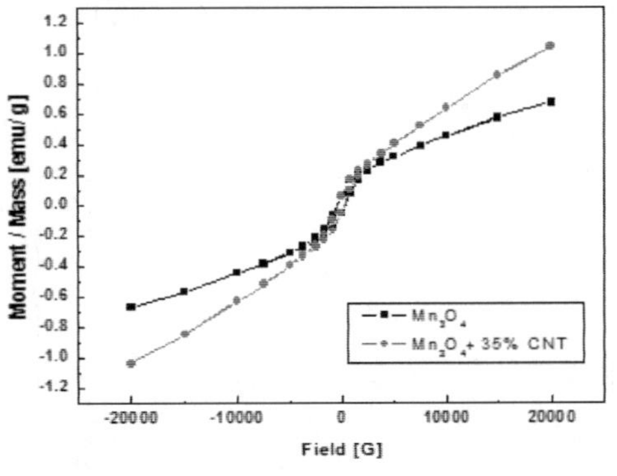

Figure 5. The relation between the magnetization and the magnetic field of the samples.

Table 1. Electrochemical Impedance parameters of of lithium cells using anodes prepared from Mn_3O_4 with and without different percentage of CNT.

Cell sample	$\mathbf{R_{ct}[\Omega]}$	σ_ω $\mathbf{[\Omega.s^{0.5}]}$	$\mathbf{D[\ cm^2s^{-1}]}$	**i° [A]**	$\mathbf{\mathit{C_{dl}}\ [F]}$
Mn_3O_4	6095.71	8182.5	5.29E-14	4.22E-06	1.64E-07
5% CNT	1872.9	1984.51	8.99E-13	1.37E-05	4.26E-10
15% CNT	1545.17	1841.95	1.04E-12	1.66E-05	8.18E-10
25% CNT	164.51	451.496	1.74E-11	1.56E-04	4.85E-07
35% CNT	127.0218	3.39E+02	3.09E-11	2.02E-04	6.28E-07
50% CNT	928	3.44E+02	2.99E-11	2.77E-05	1.36E-09

The straight lines are attributed to the diffusion of the lithium ions into the bulk of the electrode material, or the so-called Warburg diffusion. This relation is governed by equation 2. It is observed that the Warburg impedance coefficient, σ_w is 339 $\Omega.s^{0.5}$ for cell composed from Mn_2O_3 with 35% CNT cell, and it is the lowest value among the other cells. The parameters of the electrochemical impedance spectroscopy are presented in Table 1. Also, the diffusion coefficient values of the lithium ions for diffusion into the bulk electrode materials have been calculated using Equation (3) and are recorded in Table 1.

$$Z_{re} = R_e + R_{ct} + \sigma_W.\omega^{-0.5} \quad (2)$$

$$D = 0.5(\frac{RT}{A\ F2\sigma w\ C})^2 \quad (3)$$

$$Z_{re} = R_e + R_{ct} + 2\sigma_W^{\ 2}.\ C_{dl} \quad (4)$$

The double layer capacitance is given by equation (5).

$$\omega\ = 1 / R_{ct}.C_{dl} \quad (5)$$

where: R_{ct}: charge transfer resistance, R_e: electrolyte resistance, ω: angular frequency in low frequency region, D: diffusion coefficient, R: the gas constant, T: the absolute temperature, F: Faraday's constant, A: the area of electrode surface, and C: molar concentration of Li^+ ions [2,15]. The obtained diffusion coefficient, $3.09x10^{-13}$ cm^2s^{-1} for Mn_3O_4 with 35% CNT cell explains the higher mobility for Li^+ ion diffusion than in the other cell. Furthermore, the exchange current density is given by the following formula:i° = RT/nFR_{ct}, where n is the number of electron involved in the electrochemical reaction. It is observed that C_{dl} for cell " $Mn_3O_4$with 35% CNT " has the highest value: 1.66 x 10^{-8} F.

The magnetic properties of Mn_3O_4 and Mn_3O_4 with 35% CNT samples are shown in Figure 5. It is observed that the magnetization of Mn3O4 with 35% CNT sample is higher than Mn_3O_4 one. The magnetic coercivities (Hc) are 322.98 and 365.39 G for samples without CNT and with 35% CNT, respectively as shown in Figure 5. This means that the presence of 35% C increases the H_c. i.e. Mn_3O_4 with 35% CNT is more ferromagnetic than Mn_3O_4 itself.

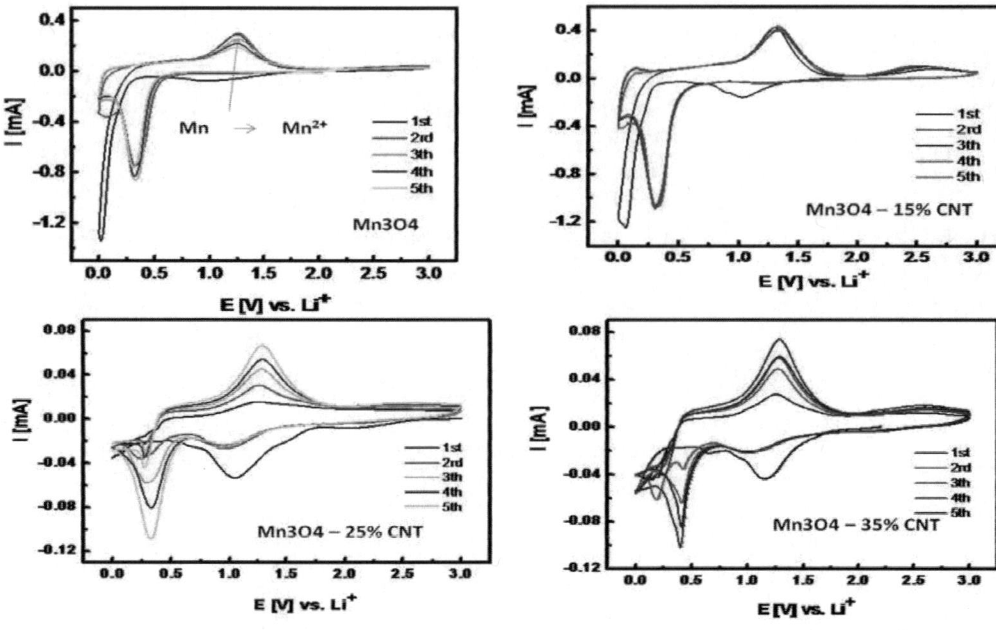

Figure 6. Cyclic voltammograms of Li/ Mn_3O_4-CNT cells.

3.3 Potentiodynamic and galvanostatic measurements

Cyclic voltammetric (CV) measurements were carried out between 0 and 3 V as shown in Figure 6. The cyclic voltammogram of the investigated Mn_3O_4 sample shows two cathodic reduction peaks at 0.9 (broad) and 0.3 V vs. Li+. The first peak for formation of a solid electrolyte interphase (SEI) film owing to the decomposition of electrolyte on the Mn_3O_4 particle surface and reduction of Mn^{3+} to Mn^{2+} [8], while the second one (0. 3V) for Mn^{2+} to Mn(0) and the formation of amorphous Li_2O based on the electrochemical conversion reaction as follows:
On the other hand, the anodic oxidation peaks take place at 1.3 and 2.6V vs. Li+. These peaks are attributed to the oxidation of Mn^0 to Mn^{2+} and the other is assigned to the re–oxidization of Mn^{2+} to higher oxidation state manganese [13]. It is observed that the addition of the CNT with 25 or 35% increases the current intensities of the peaks.
The first discharge/charge capacity plateaus versus the working voltage between 3 and 0.01V vs. Li^+ are shown in Figure 7. The first discharge capacities of the cells are nearly about 1050 $mAhg^{-1}$. The profiles for the first discharge curves look fairly similar for the investigated cells. There is a small transient and slope plateau between 1.5 and 0.25V vs. Li^+ for the formation of solid electrolyte interface (SEI) film and the reduction of Mn_3O_4 to Mn^{2+}, respectively. A plateau at 0.25 V is observed that corresponded to further reduction of Mn^{2+} to Mn(0) [7–9,16].
Figure 8 shows the curves of the specific discharge capacity vs. the cycle number of the different cells samples in the discharge voltage range of 3 and 0.01 V. For Li/ Mn_3O_4- 35% CNT cell, the capacity can be retained to about 90% of the discharge capacity after 100 cycles with respect to 2nd cycle and the specific discharge capacity is about 688 $mAhg^{-1}$, which is the highest value among the other cells. The initial capacity loss is attributed to the surface film formation, which is known as solid electrolyte interface (SEI) film. This is due the formation of irreversible reaction as: $8Li^+ + 8e^- + Mn_3O_4 \rightarrow 4Li_2O$

+ 3Mn [17,18]. It is observed from CV measurements (Figure 6) that the 2nd and the followed cycles of Mn_3O_4 have no cathodic reduction peaks at 1V butMn_3O_4 with 25 and 35% CNT have this type peak. This means that quasi-reversible of Mn_3O_4/CNT composite is formed as reported with similar Mn_3O_4/Graphite composite [17]. It can be explained that the reaction is an intercalation reaction when the discharge voltage was limited at 1.5V, the frame structure of material almost was not destroyed, and however, the reaction was a probable redox reaction when the discharge voltage went to lower potential (i.e. 1.5–0.01 V), then the structure of the materials is entirely destroyed at very low potential and cannot be recovered. It can be expected that the charge/discharge voltage is the key factor to affect the cyclic performance.

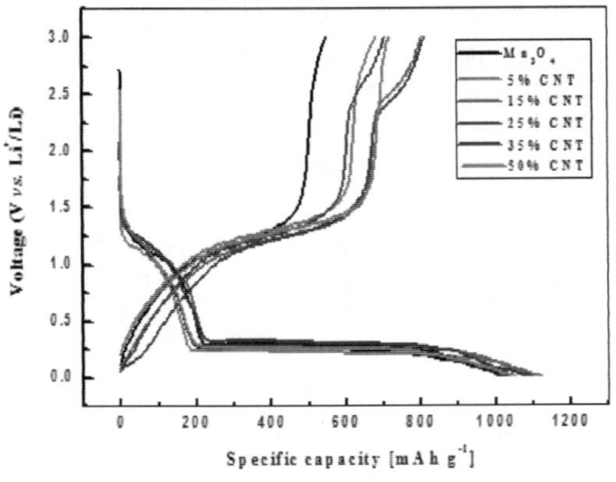

Figure 7. Capacity- voltage profile of Li/ Mn_3O_4-CNT cells.

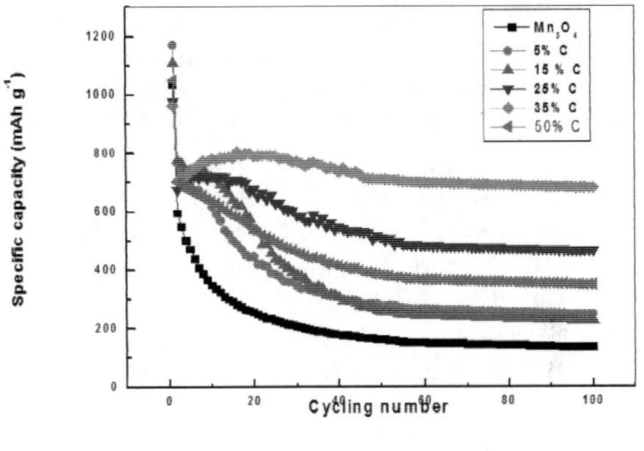

Figure 8. Cycle life performance of Li/ Mn_3O_4-CNT cells.

4. Conclusion

Different ratios of Mn_3O_4/CNT (carbon nano tubes) samples were prepared by hydrothermal reaction. Their crystal structures of the samples are tetragonal system. The cell of Mn_3O_4/35% CNT has lowest charge transfer resistance, R_{ct}, 19.18 Ω. The magnetic coercivities (H_c) are 322.98 and 365.39 G for samples without CNT and with 35% CNT, respectively. For Li/ Mn_3O_4- 35% CNT cell, the capacity can be retained to about 90% of the discharge capacity after 100 cycles and the specific discharge capacity is about 688 $mAhg^{-1}$, which is the highest among the other cells at discharge current density $0.02Acm^{-2}$.

Acknowledgment

The author would like to thank Prof. Dr. Jiazhao Wang and Dr. HaipengGuo, Institute for Superconducting and Electronic Materials, University of Wollongong, Australia for their assistance in this work.

References

[1] Wenelska K, Ottmann A, Schneider P, Thauer E, Klingeler R and Mijowska E 2016 Hollow carbon sphere/metal oxide nanocomposites anodes for lithium-ion batteries *Energy***103** 100–6
[2] Shenouda A Y and Liu H K 2010 Preparation, characterization, and electrochemical performance of Li2CuSnO4 and Li2CuSnSiO6 electrodes for lithium batteries *J. Electrochem. Soc.***157** A1183–7
[3] Shenouda A Y and Liu H K 2008 Electrochemical behaviour of tin borophosphate negative electrodes for energy storage systems *J. Power Sources***185** 1386–91

[4] Gangaraju D, Sridhar V, Lee I and Park H 2017 Graphene-carbon nanotube-Mn3O4 mesoporous nano-alloys as high capacity anodes for lithium-ion batteries *J. Alloys Compd.***699** 106–11
[5] Liang M and Zhi L 2009 Graphene-based electrode materials for rechargeable lithium batteries *J. Mater. Chem.***19** 5871–8
[6] Wang Z-H, Yuan L-X, Shao Q-G, Huang F and Huang Y-H 2012 Mn3O4 nanocrystals anchored on multi-walled carbon nanotubes as high-performance anode materials for lithium-ion batteries *Mater. Lett.***80** 110–3
[7] Yang Z, Lu D, Zhao R, Gao A and Chen H 2017 Synthesis of a novel structured Mn3O4@ C composite and its performance as anode for lithium ion battery *Mater. Lett.***198** 97–100
[8] Jing M, Hou H, Yang Y, Zhang Y, Yang X, Chen Q and Ji X 2015 Electrochemically alternating voltage induced Mn3O4/graphite powder composite with enhanced electrochemical performances for lithium-ion batteries *Electrochim. Acta***155** 157–63
[9] Luo S, Wu H, Wu Y, Jiang K, Wang J and Fan S 2014 Mn3O4 nanoparticles anchored on continuous carbon nanotube network as superior anodes for lithium ion batteries *J. Power Sources***249** 463–9
[10] Xu S-D, Zhu Y-B, Zhuang Q-C and Wu C 2013 Hydrothermal synthesis of manganese oxides/carbon nanotubes composites as anode materials for lithium ion batteries *Mater. Res. Bull.***48** 3479–84
[11] Park S-H and Lee W-J 2015 Hierarchically mesoporous carbon nanofiber/Mn3O4 coaxial nanocables as anodes in lithium ion batteries *J. Power Sources***281** 301–9
[12] Zhao Y, Ma C and Li Y 2017 One-step microwave preparation of a Mn3O4 nanoparticles/exfoliated graphite composite as superior anode materials for Li-ion batteries *Chem. Phys. Lett.***673** 19–23
[13] Cui X, Wang Y, Chen Z, Zhou H, Xu Q, Sun P, Zhou J, Xia L, Sun Y and Lu Y 2015 Preparation of pompon-like MnO/carbon nanotube composite microspheres as anodes for lithium ion batteries *Electrochim. Acta***180** 858–65
[14] Yang X, Wang X, Zhang G, Zheng J, Wang T, Liu X, Shu C, Jiang L and Wang C 2012 Enhanced electrocatalytic performance for methanol oxidation of Pt nanoparticles on Mn3O4-modified multi-walled carbon nanotubes *Int. J. Hydrogen Energy* **37** 11167–75
[15] Faulkner A J B and L R*Electrochemical Methods* (John Wiley & Sons, New York.)
[16] Fang X, Lu X, Guo X, Mao Y, Hu Y-S, Wang J, Wang Z, Wu F, Liu H and Chen L 2010 Electrode reactions of manganese oxides for secondary lithium batteries *Electrochem. commun.***12** 1520–3
[17] Hao Q, Wang J and Xu C 2014 Facile preparation of Mn 3 O 4 octahedra and their long-term cycle life as an anode material for Li-ion batteries *J. Mater. Chem. A* **2** 87–93
[18] Zhuang Y, Ma Z, Deng Y, Song X, Zuo X, Xiao X and Nan J 2017 Sandwich-like Mn3O4/carbon nanofragment composites with a higher capacity than commercial graphite and hierarchical voltage plateaus for lithium ion batteries *Electrochim. Acta* **245** 448–55.

Towards optimizing the conventional and ultrasonic-assisted extraction of sulfated polysaccharides from marine algae

Hanaa Essa [1,2], Daisy Fleita [1], Dalia Rifaat [1], Shady Samy [1], Mayyada El-Sayed [1,3*]

[1] Chemistry Department, American University in Cairo, AUC Avenue, New Cairo 11835, Egypt
[2] Pesticides Phytotoxicity Department, Agriculture Research Centre, 7 Shooting club St., Dokki, Giza, Egypt
[3] Chemical Engineering Department, National Research Centre, Al Bohooth St., Dokki, Giza, 12622, Egypt

Abstract. Recently, there has been a growing interest in sulfated polysaccharides extracted from various natural sources due to their versatile biological activities that find great use in biomedical and industrial applications. In the present study, sulfated polysaccharides were extracted from green algae, *Ulva lactuca* and red algae, *Jania rubens* collected from Alexandria city coast, Egypt. Conventional and ultrasonic-assisted extraction methods were investigated under different operating parameters and the optimum conditions for extraction were obtained based on the yield. Algal extracts with optimum SPs yields were then tested for their phenolic and sugar contents, as well as their 1, 1-Diphenyl-2-picrylhydrazyl (DPPH) antioxidant activities. Optimum yields of both green and red algae were obtained via extraction under ultrasonic conditions for 4h. Higher yields were produced from the green algae as compared to the red ones. However, red algal extracts showed relatively higher antioxidant activities, possibly due to their higher mannose content.

***Keywords*:** *Ulva lactuca*, *Jania rubens*, marine algae, antioxidant activity, phenolic content

1. Introduction

Marine algae are rich in various potentially bioactive compounds, a major group of which are sulfated polysaccharides (SPs) which are anionic carbohydrate biopolymers consisting of repeated monosaccharide units linked by glycosidic bonds. Algal SPs have drawn great attention in the biomedical as well as food and pharmaceutical industrial areas because of their broad spectrum of biological activities including anticoagulation, antiviral, anti-tumor, anti-inflammation and antioxidation. The potential of marine algae, being green (Chlorophyceae), red (Rhodophyceae) or brown (Phaeophyceae), as a source of bioactive compounds has not yet been fully explored especially with the increase in the health awareness and the advancement in functional foods and pharmaceutical industries [1–6]. Antioxidant activity, in particular, has become a subject of intensive investigations due to the growing demand to develop natural anti-aging and anti-carcinogenic compounds with health benefits. However, the search for natural antioxidants as alternatives to synthetic ones is of great interest due to safety and cytotoxicity issues associated with the latter [7–10].

Aqueous extraction of the green algae *Ulva lactuca* produced SPs that were mainly constituting L-rhamnose, D-xylose and D-glucose in the molar proportion of 4.2: 1.3: 1.0, in addition to traces of D-mannose, 24% of D-glucuronic acid and 19% of ester sulfate [11]. Furthermore, acid-hydrolysis revealed the presence of the following monosaccharide components: arabinose, xylose, rhamnose, galactose, mannose and glucose in the ratio of 1: 1: 9: 5: 2.5: 16, respectively. As for the red sea weeds, sulfated galactans are its most abundant constituents. The main acidic polysaccharides from the red seaweed *Jania rubens* share the general characteristics of corallinans (agar-like xylogalactans) [12].

The yields and biological activities of SPs produced from algae depend not only on the type of algae, but on the underlying extraction method and the relevant operating conditions. In this regard, a number of conventional techniques for extraction have been reported. The most commonly used ones were employing water [13–19]and other solvents [20–32]. This technique, however, suffers from being time-consuming and hence energy intensive and sometimes not eco-friendly especially when extraction involves the excessive use of organic solvents. Therefore, several modern techniques were proposed in an attempt to overcome these drawbacks. One example of these techniques is the enzyme-assisted extraction (EAE) which has been reported to increase yields and reduce solvent consumption [33–35]. However, the method limitations include high cost of enzymes and unavailability of substrate-specific enzymes. The applications of other relatively recent techniques such as pressurized liquid extraction (PLE) and supercritical fluid extraction (SFE) were also very limited due to their high energy requirements and subsequently high cost [36]. Furthermore, the high temperature involved in microwave-assisted extraction (MAE) and SFE could be an issue for algal species that are thermolabile. Thus with the increasing energy costs and the drive to reduce greenhouse gas emissions, food and plant- based chemical industries are thriving to come up with new technologies that provide shorter extraction times and reduced organic solvent consumption which, in turn, would save energy and cost while increasing quality and safety. This has led to the consideration of the use of enhanced and efficient extraction techniques such as ultrasound-assisted extraction (UAE).

Advances in UAE have resulted in a number of innovative techniques such as ultrasound-assisted Soxhlet apparatus, ultrasound-assisted Clevenger distillation, continuous ultrasound- assisted extraction and combination of ultrasound with other techniques such as microwave extrusion and supercritical fluid extraction [37]. In recent years, UAE of antioxidants has been widely applied due to its high efficiency, reproducibility and ease of operation. The force of ultrasound can produce heat and disrupt cell walls, thus facilitating the release of bioactive compounds. However, it was claimed that although this promising technique has presented numerous advantages over conventional techniques, yet evidences revealed that it caused degradation of some compounds and modifications in the physicochemical properties of products [38]. Very few studies investigated the use of modern techniques such as UAE [39] and MAE [40] in order to increase the yield and reduce the extraction time.

In the current study, conventional and ultrasonic-assisted extraction methods will be investigated and comparatively assessed under different operating conditions and using red and green algae in order to select a viable efficient extraction method that produces optimum SPs yields with potent antioxidant activities. This comparative screening approach for scouting the optimum extraction conditions should be of great benefit in the design and development of efficient algal extraction processes for food and pharmaceutical industries.

2. Materials and Methods

2.1 Collection of Seaweed Samples

Ulva lactuca and *Jania rubens* were collected from Alexandria city coast (N 31° 19` E030° 03`) and were immediately brought to the laboratory in sterile plastic bags containing seawater in order to prevent evaporation. The algal samples were first cleaned well with the seawater until unwanted impurities, adhering sand particles and epiphytes, pebbles, and shells were removed. They were then washed thoroughly with tap water followed by deionized water to remove the surface salty materials. Afterwards, algae were air dried for one week then finely- ground in an electric mixer. The powdered samples were subsequently stored in the refrigerator for future use.

2.2 Extraction of Seaweeds

2.2.1 Conventional extraction. Ten grams of each of the prepared powdered seaweed samples were suspended in 200 mL of distilled water. The suspension was agitated on a magnetic stirrer and its pH was adjusted to 5.5 using1N HCl. It was then shaken either for 24 h at room temperature (RT) or for 2 h at 80 ^{0}C, and subsequently filtered through cheese cloth. The filtrate was dialyzed overnight in a dialysis bag, after which it was centrifuged (Heraeus-Christ, GMBH336 Osteode Ma Harz No.39189) for 10 min at -10 ^{0}C and 5500-6000 rpm. The supernatant was collected and added to ethanol in the ratio of 1:3.5 (v/v), respectively. After further centrifugation under the aforementioned conditions, the supernatant (SP fraction) was collected.

2.2.2 Ultrasonic-assisted extraction. Ultrasonic-assisted experiments were conducted for different periods of time in the ultrasonicator (VWR® sonicator model 150HT) at 50 and 60°C for *Ulva lactuca* and at RT for *Jania rubens*. Samples were prepared and SP fractions were extracted in the same manner as in the conventional extraction.

2.3 Characterization and chemical analysis of the algal extracts

2.3.1 Chemical analysis. **Total carbohydrate content** of the polysaccharide extracts was determined by the phenol-sulfuric method [41] using galactose as standard, while the **protein content** was determined adopting the method of Lowery [42] and using bovine albumin as standard. **Total phenolic content** of polysaccharide fractions was determined using the Folin-Ciocalteu reagent according to the method described by Singleten & Rossi [43], with minor modifications.

2.3.2 IR Spectroscopy analysis. The IR spectra of the extracts were determined using a Fourier transform IR spectrophotometer (FT-IR) (TGA/FT-IR Nicolet 380). Samples were mixed with KBr powder, ground and then pressed into 1-mm pellets for measurement in the frequency range of 4000-500 cm^{-1}.

2.3.3 Testing antioxidant activity of the algal extracts. The free radical scavenging activity of the investigated algal extracts was measured using 1,1-Diphenyl-2-picrylhydrazyl (DPPH) antioxidant assay following the method of Blois [44] with minor modifications.

$$\text{Scavenging rate } (\%) = [A_0-(A_2-A_1)]/A_0 * 100 \qquad (1)$$

2.3.4 HPLC analysis of sugar content. Before analysis, samples were filtered through a 0.45 μm membrane. Sugar content in the filtrate was analyzed using HPLC, Shimadzu Class-VPV 5.03 (Kyoto, Japan) equipped with refractive index RID-10A, Shimadzu detector, LC-16ADVP binary pump, DCou-14 A degasser and ShodexPL Hi-PlexPb column (Sc 1011 No. H706081), Guard column Sc-LcShodex, and heater set at 80 ^{0}C. Separation and quantitation were carried out on an amino-bonded column with a mobile phase of (80:20) H_2O to acetonitrile at a flow rate of 1mL/min.

3. Results and discussion

3.1 Yields of algal extracts

Extraction of SPs from the two employed algal species was performed using both ultrasonic-assisted and conventional methods under different operating conditions. Obtained yields and antioxidant activities were then compared in order to obtain the optimum conditions of extraction pertaining to each algal species. The UAE of SPs was first investigated for the green algae at 50 and 60 ^{0}C and for the red one at RT. These temperatures were chosen because the algae were sensitive and susceptible to degradation at higher temperatures [45]. The effect of extraction time on the percentage yields of SPs (by weight) extracted from *Ulva lactuca* using the UAE method at 50 and 60 ^{0}C and different incubation periods is depicted in Figure 1.

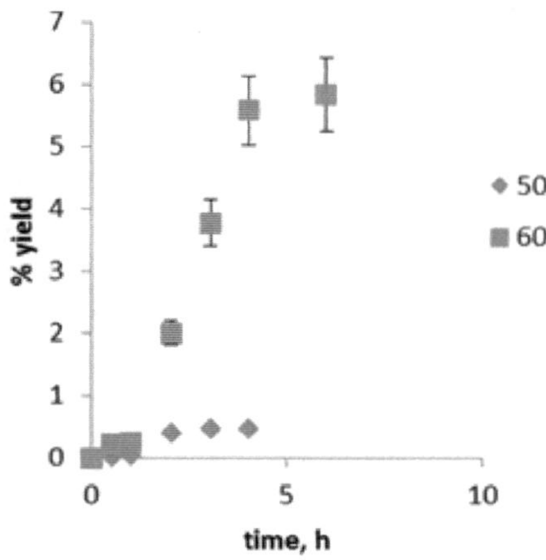

Figure 1. Time courses for the % yield of *Ulva lactuca* SPs at different temperatures.

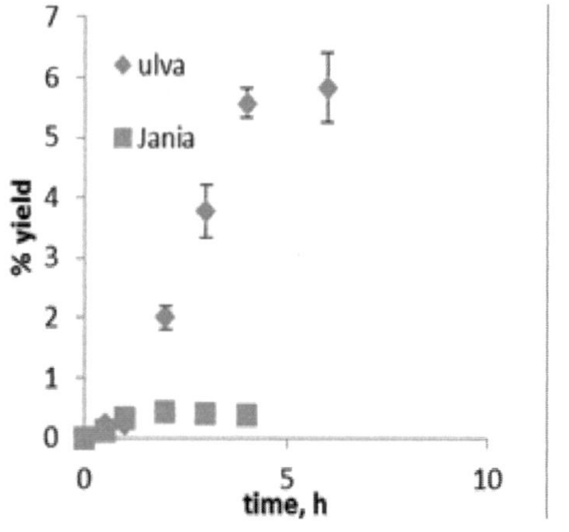

Figure 2. % Yield as a function of time for UAE SPs of *Ulva lactuca* at 60 ^{0}C and *Jania rubens* at RT.

As can be seen from the figure, the yield at both temperatures increases with prolonging the incubation time until it reaches a plateau where it remains constant. However at the higher temperature, the yield takes longer time to reach the plateau. Furthermore, yields obtained at 60 ^{0}C dramatically exceed those obtained at 50 ^{0}C which indicates that temperature is favorable for the extraction of SPs from *Ulva lactuca*. The maximum % yield obtained at 60 ^{0}C is more than ten times higher than its counterpart obtained at the lower temperature. In view of these results, the yields of *Ulva lactuca* at 60 ^{0}C and different extraction times were compared to those of *Jania rubens* at RT as shown in Figure 2. Clearly, the yields obtained from the green algae are more than 12 fold higher than those of the red algae. Maximum yields for *Ulva lactuca* and *Jania rubens* were obtained at 4 and 2 h, respectively.From the above, it can thus be deduced that the optimum yields for UAE of SPs from green algae were obtained at 60 ^{0}C and 4h extraction time, while those from red algae were obtained at RT and 2h.

Yields of SPs from ultrasonic-assisted extraction were then compared to their counterparts obtained under conventional conditions as depicted in Figure 3. For the red algae, conventional extraction was conducted at RT for 24 h; while for the green algae, two sets of operating conditions were employed. Extraction conditions for the first set are RT and 24 h, while those for the second set are 80 ^{0}C and 2h.

The figure shows that the yields of *Ulva lactuca* are much higher than those of *Jania rubens*. For the green algae, the highest yield is obtained via conventional extraction at RT and 24 h. This yield exceeds those obtained from the other adopted conventional and ultrasonic-assisted extractions by about 1.4 and 1.5 fold, respectively. In addition, the yields obtained via these other adopted methods are comparable. However, the difference between the highest yield obtained conventionally at room temperature and 24 h and the lowest yield obtained ultrasonically at 60 ^{0}C and 4 h is slightly above 2%. This finding therefore does not justify performing the extraction conventionally for 24 h to gain only a 2% increase in yield. Thus among the adopted operating conditions, UAE at 60 ^{0}C and 4 h would offer a good compromise for saving both time and energy. As for *Jania rubens*, conventional and ultrasonic-assisted extraction gave comparable yields. Again for the purpose of saving time, UAE would be more favorable than the conventional extraction.

3.2 DPPH Antioxidant activities of the SP fractions

The DPPH antioxidant activities were then measured for the algal extracts obtained under conditions that gave maximum yields. Figures 4A and B show the % DPPH scavenging activity as a function of concentration for the polysaccharide fractions extracted respectively from *Ulva lactuca* and *Jania rubens* at different extraction conditions. The inset of the figure shows the % DPPH scavenging activity for ascorbic acid which was used as a standard reference. As clear from the figures, the scavenging activity increases linearly with time. It increases by about 5-10% for *Ulva lactuca* (Figure 4A) and by about 10-20% for *Jania rubens* (Figure 4B) as the concentration increases by ten folds from 100 to 1000 μg/mL.

As shown earlier, maximum scavenging activities for all SPs extracts under the different employed extraction conditions were obtained at a concentration of 1000 μg/mL. Values thereof are presented in Figure 5.

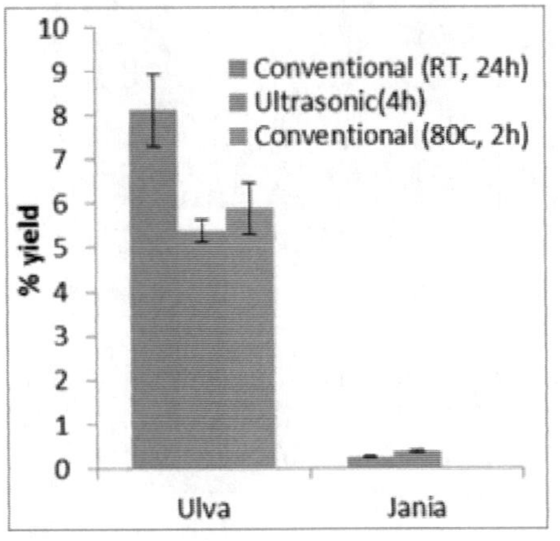

Figure 3. Yields for the green and red algal SPs obtained under optimum conventional and ultrasonic conditions. Ultrasonic temperature employed was 60 0C for Ulva lactuca and RT for Jania rubens.

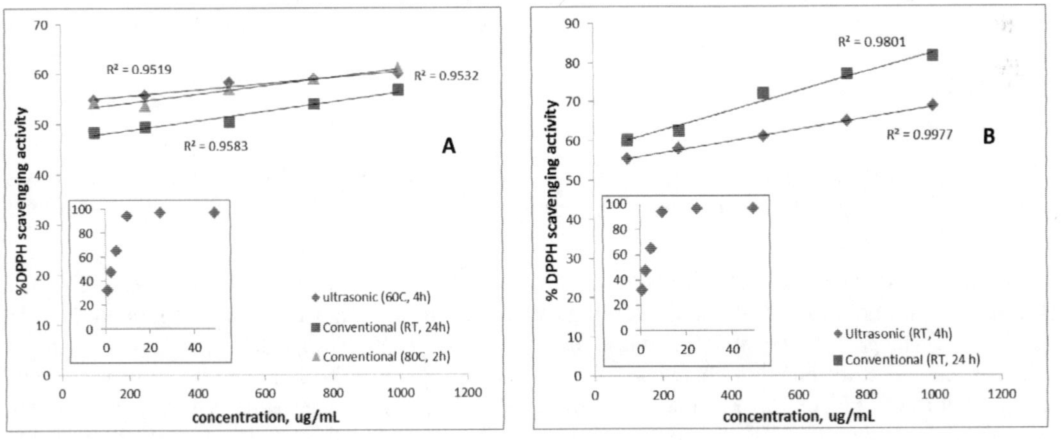

Figure 4. % DPPH scavenging activity as a function of concentration for the polysaccharide fractions extracted from *Ulva lactuca* (A) and *Jania rubens* (B) at optimum extraction conditions. The inset of the figure represents the % DPPH scavenging activity for ascorbic acid.

As clear from the figure, both the conventional and ultrasonic-assisted extraction methods yielded comparable antioxidant activities for the red algal species. The figure also shows that UAE of SPs from *Ulva lactuca* provided either slightly higher or comparable % scavenging activities to those obtained via the conventional method.

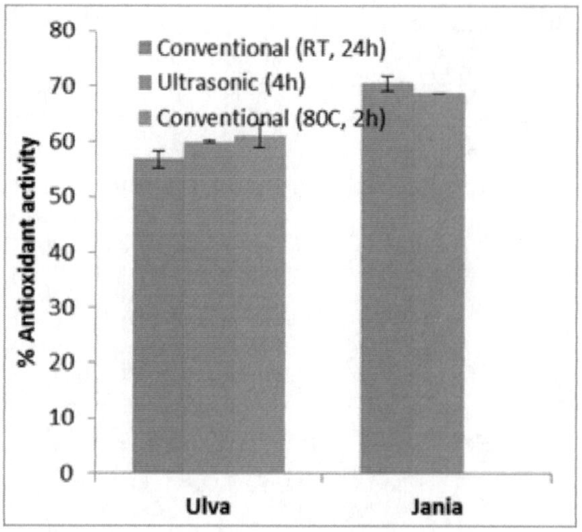

Figure 5. Maximum antioxidant activities measured using DPPH assay, for 1000 μg/mL of the different SPs extracts that were obtained under conventional and ultrasonic conditions.

3.3 Protein, carbohydrate and phenolic contents

Table 1 compiles the values of the protein, carbohydrate and phenolic contents of the algal extracts obtained at the optimum extraction conditions deduced earlier. The table also shows the corresponding percent yields and DPPH scavenging activity. For each algal extract, the higher antioxidant activities correspond to higher phenolic and protein contents probably due to the presence of hydroxyl and amide functional groups, respectively. The direct proportionality between phenolic content and scavenging activity was confirmed in previous literature [46,47]. Furthermore, the carbohydrate contents for both red and algal extracts obtained from the conventional extraction are higher than their counterparts obtained from UAE, probably due to the breaking of the carbohydrate chains by means of the ultrasonic force.

3.4 FTIR analysis

FTIR spectra of SPs extracted from *Ulva lactuca* under the previously selected optimum conditions are shown in Figure 6A. The main bands appearing in the figure and their corresponding functional groups are compiled in Table 2. These bands are characteristic of SPs functional groups [49, 50].

On the other hand, infrared spectra for the SPs extracted from *Jania rubens* are illustrated in Figure 6B. Bands at around 3400-3500 $cm^{-1}$are ascribed to vibration of O-H, while those at around 2925.8-2927.8 cm^{-1} correspond to the stretching vibration of the C-H group [6,48,49]. In addition, the band at 875.3 cm^{-1} could be attributed to galactose sulfate groups [47]. It can be observed that all these bands were present in the SPs extracted from *Ulva lactuca*. In addition, the ester sulfate band is not shown, however the presence of sulfate is confirmed by the appearance of the bands corresponding to the galactose sulfate groups.

The type of sugars present in the algal extracts and their content could affect their antioxidant activity. The molar composition of some of the sugars present in the different investigated extracts was determined using HPLC analysis and the resulting sugar contents are presented in Table 3. For *Ulva lactuca* extracts, it is clear that, among the tested sugars, glucose and galactose are the most dominant sugars in the fractions that were conventionally extracted at RT. However, glucuronic acid was the most dominant sugar in the fractions extracted conventionally at 80 ^{0}C or ultrasonically at 60 ^{0}C. High temperatures could have favored the oxidation of glucose and galactose sugars to glucuronic acid. The

IOP Conf. Series: Materials Science and Engineering **464** (2018) 012007 doi:10.1088/1757-899X/464/1/012007

higher antioxidant activities exhibited for the fractions extracted at high temperatures relative to the fraction extracted at RT could be attributed to the high glucuronic acid content. As for *Jania rubens* extracts, glucose and galactose are the dominant sugars for the ultrasonically and conventionally extracted fractions, respectively. In previous work, it was reported that conventionally extracted fractions of *Jania rubens* primarily constitute galactose which is confirmed in this work [48]. The dominant glucose content for the ultrasonically extracted fraction could be a result of the ultrasonic-aided conversion of galactose to glucose. Although glucose and galactose could be partially responsible for the antioxidant activity of *Jania rubens* extracts, the higher antioxidant activity of these extracts relative to those of the green algae could be owed to their relatively higher mannose content. We showed in previous work that mannose was responsible for the high antioxidant activity of the viscozyme hydrolysate of the *Pterocladia capillacea* extract [47].

Table 1. Protein, carbohydrate and total phenolic contents for the algal extracts obtained under optimum extraction conditions.

Algal extract	%Yield	%Protein	%Carbohydrates	Total phenolic content (mgGA/g)	%DPPH activity
Ulva.CV (RT, 24h)	8.00±0.82	4.0±0.9	34.5±3.1	36.2±2.9	50-52
Ulva. CV (80C, 2h)	5.80±0.59	7.7±0.4	26.8±2.7	71.3±6.2	53-61
Ulva.US (60C, 4h)	5.30±0.25	5.8±0.7	7.7±0.1	60.6±7.9	55-60
Jania.US (RT, 4h)	0.36±0.04	9.4±1.7	2.6±0.1	132.6±2.5	55-68
Jania.CV (RT, 24h)	0.25±0.02	19.6±3.7	8.6±0.3	136.1±0.7	64-71

Values are expressed as mean±SD (n=3)

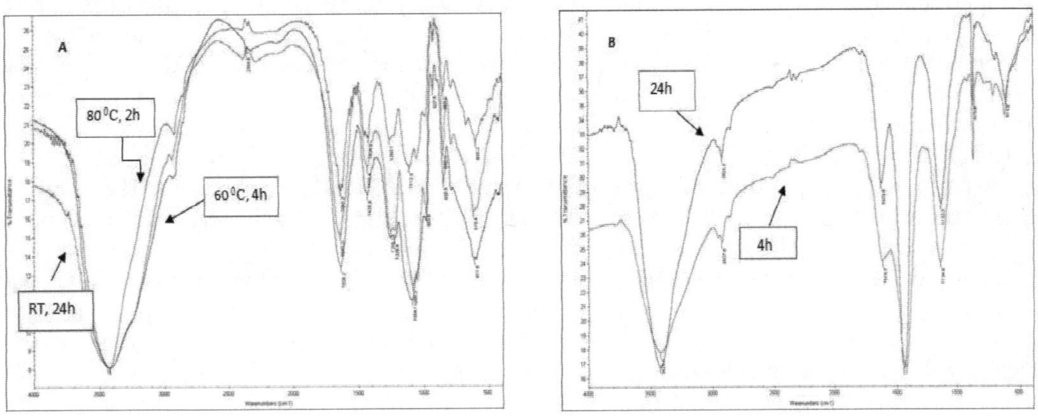

Figure 6. FTIR spectra of the extracts of **(A)** *Ulva lactuca* under conventional conditions of RT and 24 h (red), 80°C and 2 h (violet) and ultrasonic conditions of 60 °C and 4 h (blue), **(B)** *Jania rubens* under conventional conditions of RT and 24h (blue), and ultrasonic conditions of RT and 4 h (red).

Table 2. FTIR bands pertaining to Figure 6 and their corresponding functional groups

Wavelength, cm^{-1}	functional groups
3500-3400	OH groups
1433.4-1424.2	Uronic acid and phenolic groups
1260.1-1258.1	Ester Sulfate group
1088.3-112.9	Acidic polysaccharide
963.6-927.6	Glycosidic linkage
850.3-850.2	Galactose sulfate group

Table 3. Sugar content of the algal extracts, expressed as molar composition (% mol)

	Glucuronic acid	Mannose	Galactose	Glucose	Xylose
Ulva. CV (RT,24h)	8.04	0.00	29.02	62.95	0.00
Ulva. CV (80C, 2h)	94.83	0.36	0.51	1.24	3.06
Ulva. US (60C, 4h)	89.92	0.00	3.53	6.55	0.00
Jania. US (RT, 4h)	0.16	3.51	0.10	94.04	2.14
Jania. CV (RT, 24h)	0.27	2.19	94.02	0.80	2.72

4. Conclusion

This study investigates the conventional and ultrasonic-assisted extraction of sulfated polysaccharides from green *Ulva lactuca* and red *Jania rubens* algal species. For each method, extraction was conducted under different operating conditions of time and temperature in order to select the conditions that produce optimal yields of SPs. The extracts (SPs fractions) obtained under these optimum conditions were analyzed for their antioxidant activities as well as their carbohydrates, proteins and phenolic contents. It was found that the highest yields for *Ulva lactuca* were achieved under conventional extraction conditions of RT and 24h. Under ultrasonic-assisted conditions of 60°C and 4h, slightly lower yields were obtained. However for the sake of saving time and energy, the optimum extraction conditions were selected to be those of ultrasonic-assisted extraction. As for *Jania rubens*, SPs yields from conventional extraction at RT and 24h were comparable to those of ultrasonic-assisted extraction at RT and 4h, hence the latter was the optimum choice. DPPH antioxidant activity results showed that the scavenging activities for different concentrations of *Ulva lactuca* and *Jania rubens* extracts range between 55-60% and 55-68%, respectively. The relatively higher antioxidant activities exhibited by the red algae could be attributed to its high mannose content. Furthermore, the green algae fractions extracted at relatively higher temperatures showed better antioxidant activities owing to their high glucuronic content. Relevant phenolic content measurements indicated direct correlation between phenolic content and antioxidant activity. In addition, FTIR analysis confirmed the presence of OH, uronic acid and sulfate groups as well as glycosidic linkages typical of sulfated polysaccharide chains.

Acknowledgements

Thanks are due to the Chemistry Department at the American University in Cairo for funding this project

.References

[1] Kuprash L P 2003 Spirulina: perspectives of the application in medicine *Likars' ka Sprav.* 100–10

[2] Kandhasamy M and Arunachalam K D 2008 Evaluation of in vitro antibacterial property of seaweeds of southeast coast of India *African J. Biotechnol.* **7**

[3] Nair R, Chabhadiya R and Chanda S 2007 Marine algae: Screening for a potent antibacterial agent *J. Herb. Pharmacother.* **7** 73–86

[4] Kadam S U, Tiwari B K and O'Donnell C P 2013 Application of novel extraction technologies for bioactives from marine algae *J. Agric. Food Chem.* **61** 4667–75

[5] Sebaaly C, Karaki N, Chahine N, Evidente A, Yassine A, Habib J and Kanaan H 2012 Polysaccharides of the red algae "pterocladia" growing on the lebanese coast: Isolation, structural features with antioxidant and anticoagulant activities *J. Appl. Pharm. Sci.* **2** 001-010

[6] Yang Y, Liu D, Wu J, Chen Y and Wang S 2011 In vitro antioxidant activities of sulfated polysaccharide fractions extracted from Corallina officinalis *Int. J. Biol. Macromol.* **49** 1031–7

[7] Hettiarachchy N S, Glenn K C, Gnanasambandam R and Johnson M G 1996 Natural antioxidant extract from fenugreek (Trigonella foenumgraecum) for ground beef patties *J. Food Sci.* **61** 516–9

[8] Park P-J, Jung W-K, Nam K-S, Shahidi F and Kim S-K 2001 Purification and characterization of antioxidative peptides from protein hydrolysate of lecithin-free egg yolk *J. Am. Oil Chem. Soc.* **78** 651–6

[9] Duan X-J, Zhang W-W, Li X-M and Wang B-G 2006 Evaluation of antioxidant property of extract and fractions obtained from a red alga, Polysiphonia urceolata *Food Chem.* **95** 37–43

[10] Wijesekara I, Yoon N Y and Kim S K 2010 Phlorotannins from Ecklonia cava (Phaeophyceae): Biological activities and potential health benefits *Biofactors* **36** 408–14

[11] Barnes H 1966 Some contemporary studies in marine science

[12] Navarro D A and Stortz C A 2008 The system of xylogalactans from the red seaweed Jania rubens (Corallinales, Rhodophyta) *Carbohydr. Res.* **343** 2613–22

[13] Hernández-Garibay E, Zertuche-González J A and Pacheco-Ruíz I 2011 Isolation and chemical characterization of algal polysaccharides from the green seaweed *Ulva clathrata* (Roth) C. Agardh *J. Appl. Phycol.* **23** 537–42

[14] Zhang Z, Wang F, Wang X, Liu X, Hou Y and Zhang Q 2010 Extraction of the polysaccharides from five algae and their potential antioxidant activity in vitro *Carbohydr. Polym.* **82** 118–21

[15] Govindan S M, Thomas J, Pratheesh P T and Kurup G M 2012 Ex vivo anticoagulant activity of the polysaccharide isolated from Ulva fasciata *Int. J. Life Sci. Pharm. Res* **1** 194–7

[16] Souza B W S, Cerqueira M A, Bourbon A I, Pinheiro A C, Martins J T, Teixeira J A, Coimbra M A and Vicente A A 2012 Chemical characterization and antioxidant activity of sulfated polysaccharide from the red seaweed Gracilaria birdiae *Food Hydrocoll.* **27** 287–92

[17] Ambreen A, Hira K, Ruqqia A and Sultana V 2012 Evaluation of biochemical component and antimicrobial activity of some seaweeeds occurring at Karachi Coast *Pakistan J. Bot.* **44** 1799–803

[18] El-Baky A H, El Baz H K F and El-Latife S A 2014 Induction of sulfated polysaccharides in Spirulina platensis as response to nitrogen concentration and its biological evaluation *J. Aquac. Res. Dev.* **5** 1

[19] Zeid A H A, Aboutabl E Al, Sleem A A and El-Rafie H M 2014 Water soluble polysaccharides extracted from Pterocladia capillacea and Dictyopteris membranacea and their biological activities *Carbohydr. Polym.* **113** 62–6

[20] Wefky S and Ghobrial M 2008 Studies on the bioactivity of different solvents extracts of selected marine macroalgae against fish pathogens *Res J Microbiol* **3** 673–82

[21] Alang G, Kaur R, Singh A, Budlakoti P and Singla P 2009 Antimicrobial activity of Ulva lactuca extracts and its fractions *Pharmacol.* **3** 107–17

[22] Babu M, Johnson M, Raja D, Arockiaraj A and Vinnarasi J 2014 Chemical constituents and

their biological activity of Ulva lactuca Linn *Int. J. Pharm. Drug Anal.* **2** 595–600
[23] Salunke MA S C and W B 2014 Anticancer activity of methanol extract of Jania rubens Linn. against Ehrlich as-cites carcinoma induced Balb/C Mice*World J. Pharm. Sci.* **2** 1224-1229
[24] Pushparaj A, Raubbin R S and Balasankar T 2014 Antibacterial activity of Kappaphycus alvarezii and Ulva lactuca extracts against human pathogenic bacteria *Int J Curr Microbiol App Sci* **3** 432–6
[25] Hassan S M and Ghareib H R 2009 Bioactivity of Ulva lactuca L. acetone extract on germination and growth of lettuce and tomato plants *African J. Biotechnol.* **8**
[26] Aruna P, Mansuya P, Sridhar S, Kumar J S and Babu S 2010 Pharmacognostical and antifungal activity of selected seaweeds from Gulf of Mannar region *Recent Res. Sci. Technol.* **2**
[27] Osman M E H, Abushady A M and Elshobary M E 2010 In vitro screening of antimicrobial activity of extracts of some macroalgae collected from Abu-Qir bay Alexandria, Egypt *African J. Biotechnol.* **9** 7203–8
[28] Uma R, Sivasubramanian V and Devaraj S N 2011 Evaluation of in vitro antioxidant activities and antiproliferative activity of green microalgae, Desmococcus olivaceous and Chlorococcum humicola *J. Algal Biomass Utln* **2** 82–93
[29] N K A and S 2012 Bio-screening of some marine algae from the coasts of Egypt. *J. Sci. Res. Pharm.* **1**17-21
[30] Hemalatha A, Girija K, Parthiban C, Saranya C and Anantharaman P 2013 Antioxidant properties and total phenolic content of a marine diatom, Navicula clavata and green microalgae, Chlorella marina and Dunaliella salina *Adv. App. Sci. Res* **4** 151–7
[31] Saritha K, Mani A E, Priyalaxmi M and Patterson J 2013 Antibacterial activity and biochemical constituents of seaweed Ulva lactuca *Glob. J. Pharmacol* **7** 276–82
[32] Chihadeh W N, Elmanama A A and Elnabris K J 2013 Antibacterial activity of four marine seaweeds collected from the coast of Gaza Strip, Palestine. *Mesopotamian J. Mar. Sci.* **28** 81–92
[33] Adalbjörnsson B V. and Jónsdóttir R 2015 Enzyme-enhanced extraction of antioxidant ingredients from algae *Methods Mol. Biol.* **1308** 145–50
[34] Wijesinghe W and Jeon Y-J 2012 Enzyme-assistant extraction (EAE) of bioactive components: a useful approach for recovery of industrially important metabolites from seaweeds: a review *Fitoterapia* **83** 6–12
[35] Li S, Han D and Row K H 2012 Optimization of enzymatic extraction of polysaccharides from some marine algae by response surface methodology *Korean J. Chem. Eng.* **29** 650–6
[36] Plaza M, Amigo-Benavent M, del Castillo M D, Ibáñez E and Herrero M 2010 Facts about the formation of new antioxidants in natural samples after subcritical water extraction *Food Res. Int.* **43** 2341–8
[37] Chemat F, Rombaut N, Sicaire A-G, Meullemiestre A, Fabiano-Tixier A-S and Abert-Vian M 2017 Ultrasound assisted extraction of food and natural products. Mechanisms, techniques, combinations, protocols and applications. A review *Ultrason. Sonochem.* **34** 540–60
[38] Rostagno M A and Prado J M 2013 *Natural product extraction: principles and applications* (Royal Society of Chemistry)
[39] Shao P, Chen X and Sun P 2013 In vitro antioxidant and antitumor activities of different sulfated polysaccharides isolated from three algae *Int. J. Biol. Macromol.* **62** 155–61
[40] Rodriguez-Jasso R M, Mussatto S I, Pastrana L, Aguilar C N and Teixeira J A 2011 Microwave-assisted extraction of sulfated polysaccharides (fucoidan) from brown seaweed *Carbohydr. Polym.* **86** 1137–44
[41] Lowry O H, Rosebrough N J, Farr A L and Randall R J 1951 Protein measurement with the Folin phenol reagent *J. Biol. Chem.* **193** 265–75
[42] LARSEN B J R N, HAUG A and Painter T J 1966 Sulphated polysaccharides in brown algae *Acta chem. scand* **20** 219–30
[43] Singleton V L and Rossi J A 1965 Colorimetry of total phenolics with phosphomolybdic-

phosphotungstic acid reagents *Am. J. Enol. Vitic.* **16** 144–58
[44] Ktari L, Blond A and Guyot M 2000 16β-hydroxy-5α-cholestane-3,6-dione, a novel cytotoxic oxysterol from the red alga Jania rubens *Bioorganic Med. Chem. Lett.* **10** 2563–5
[45] Imjongjairak S, Ratanakhanokchai K, Laohakunjit N, Tachaapaikoon C, Pason P and Waeonukul R 2016 Biochemical characteristics and antioxidant activity of crude and purified sulfated polysaccharides from Gracilaria fisheri *Biosci. Biotechnol. Biochem.* **80** 524–32
[46] Fleita D, El-Sayed M and Rifaat D 2015 Evaluation of the antioxidant activity of enzymatically-hydrolyzed sulfated polysaccharides extracted from red algae; Pterocladia capillacea *LWT-Food Sci. Technol.* **63** 1236–44
[47] Ananthi S, Raghavendran H R B, Sunil A G, Gayathri V, Ramakrishnan G and Vasanthi H R 2010 In vitro antioxidant and in vivo anti-inflammatory potential of crude polysaccharide from Turbinaria ornata (Marine Brown Alga) *Food Chem. Toxicol.* **48** 187–92
[48] Percival E and Wold J K 1963 1040. The acid polysaccharide from the green seaweed Ulva lactuca. Part II. The site of the ester sulphate *J. Chem. Soc.* 5459–68
[49] El-Rafie H M, El-Rafie Mh and Zahran M K 2013 Green synthesis of silver nanoparticles using polysaccharides extracted from marine macro algae *Carbohydr. Polym.* **96** 403–10

IOP Conf. Series: Materials Science and Engineering **464** (2018) 012008 doi:10.1088/1757-899X/464/1/012008

Estimation of landfill gases in Egypt, 1- evaluation and validation of different models

Ihab El Sersy[1], Asmaa S Hamouda[2], Nabil Abdelmonem[1], Shadia Elshishini[1] and Ibrahim Ismail[3]

[1]Department of Chemical Engineering, Cairo University, Giza, Egypt.
[2]Department of Environmental Science and Industrial Development, Faculty of Postgraduates for Advanced Sciences, Beni-Suef University, Beni Suef, Egypt.
[3]Zewail City for Science and Technology, Renewable Engineering Department, Giza, Egypt.

Abstract. Alexandria landfill site in El Hammam operated by Onyx since 2003 has been used as an Egyptian landfill case study. Methane generation rates have been estimated using three different simulation models namely: LandGEM, E-PLUS, and the 2006 IPCC Waste Model. The models were run using actual waste quantities disposed at this site together with projected waste quantities till the closure of the landfill site. Power consumption becomes economically feasible only if electricity price is increased locally.

1. Introduction

Greenhouse gas emissions are of high worldwide concern due to their serious effect on climate changes. The whole mankind is responsible for its continuation, which is having significant and negative impacts on people health and life style, natural resources specially water, and economic conditions of the whole world [1–7]. The major greenhouse gases (GHGs) are H_2O (g), CO_2, CH_4 and N_2O. Their relative appendance in the emissions is H_2O (g) (36–70%), CO_2 (9–26%), CH_4 (4–9%), and N_2O (3–7%) plus other trace gases [5]. CO_2 and CH_4 are the first and second responsible gases for the global warming [8]. The emissions of these gases are caused by both natural and anthropogenic actions.

Since the preindustrial period, the atmospheric concentration of methane has increased from approximately 400 to 700 ppb up to a global average of 1774 ppb till 2005 [6]; the rate of increase is around 0.6–0.8% annually [9]. Among the different anthropogenic resources of greenhouse gases, the gases generated form Municipal Solid Waste, MSW, dumping sites, landfills, are considered the highest growing source. Landfills represent the dominant alternative for municipal solid waste (MSW) disposal in many parts of the world. For example, it is the utilized technology for 54% of MSW generated in US, 95% of MSW generated in Australia while in, almost 100% of MSW generated in South Africa and over 80% of MSW generated in China. In EU (European Union) Member States, waste management policies and the utilization of refused derivative fuel (RDF) produced from MSW in cement industry have resulted in a reduction of MSW sent to landfills from 62% in 1995 to 41% in 2007 [10]. In principle, any place where Municipal Solid Waste, MSW, is dumped or disposed of in large quantities is considered as a bioreactor generating leachate and gases [10]. The bio-reactions in the waste are dependent upon a series of conditions such as moisture content of waste, composition, availability of oxygen "redox potential", temperature, micro-flora, and compaction rate. Under strict anaerobic conditions, methane and carbon dioxide are often the primary gases generated. When generated in a landfill, this gas is often known as landfill gas, LFG. Landfill gas is an end-product of the decomposition of biodegradable wastes in a landfill site and is itself legally considered to be a

waste. During recent years and due to the extraordinary increase in domestic wastes, landfills have become important methane emission sources [11–13]. Globally, it is estimated that more than 10% of the total anthropogenic methane emissions originate from landfill gases [14]. Moreover, recent results have shown that landfills generate up to 30% of anthropogenic methane in Europe [15].
Although it is well understood and widely accepted that the impact of LFG emissions on global warming is unignorable, still its direct effect on urban communities in the vicinity of the landfill, which are directly exposed to its emissions, is of great concern. The emissions could cause negative effects when migrating offsite, e.g. damage to vegetation or even explosion hazards. Therefore measuring, estimating and controlling of methane emission on landfills is of great importance [11]. Besides, the estimation of LFG amount is an important process to evaluate the possibility of utilizing this gas in an environmental and economical way. Several reports were published in recent years proposing or utilizing different models for such purpose [8,12,14–24]. The landfill sites and LFG management was the topic of different reports trying to propose the optimum performance and management processes for both of the landfills themselves and LFG generated [25–28].

2. Materials and methods

2.1 Landfill Site

The landfill site at El Hammam has been selected as case study as it fulfills the following requirements for commercial recovery of generated LFG. The landfill site should receive at least 200 ton/day of waste, be designed for a minimum total capacity of 2.5 million ton, and have a minimum filling height of 10 meters. Furthermore, the waste should not have been deposited for more than 5 - 10 years before LFG recovery is attempted. El Hamman landfill site could receive a total of 9.424 million tons and its lifetime is expected to extend up to year 2026 as shown in Table 1. The data provided by the landfill management (Onyx Co) Onyx Company shows that 50% of the waste is organic matter. LFG generation is due to biodegradation of organic waste which follows a pattern of five phases, with different micro-organisms involved at each stage.

Table 1. Requirement for LFG recovery project

	Minimum Requirement	**El Hammam Landfill Site (Pelt R., 1998)**
Waste Quantity (t/d)	200	2,600
Total capacity (million ton)	2.5	9.424
Filling height (m)	10	11
Operation start date (y)	5-10	5

Phase 1: Hydrolysis and aerobic degradation by aerobic micro-biota which metabolize a fraction of the organic waste to produce simpler hydrocarbons, water and carbon dioxide, which is an exothermic process.

Phase 2: Hydrolysis and Fermentation (Acidogenic), which is an aerobic process with the development of anaerobic conditions. A fermentation process takes place developing acids in the leachate.

Phase 3: Acetogenic (Acetogenesis) where anaerobic conditions are established and methanogenic conditions emerge. The LFG will start to contain increasing quantities of methane [11].

Phase 4: Methanogenic where methane production is at its highest, with a stable concentration of 40-60% CH4 by volume. At this point, the chemical processes involved are comparatively slow and can take many years to complete.

Phase 5: Oxidation where methane production will begin to decrease and the presence of atmospheric air will reintroduce aerobic conditions. This phase may be reached only after many decades.

2.2 Approach to achieve study objectives

Figure 1 presents the approach followed to achieve the study objectives of selecting the most cost-effective option of energy utilization.

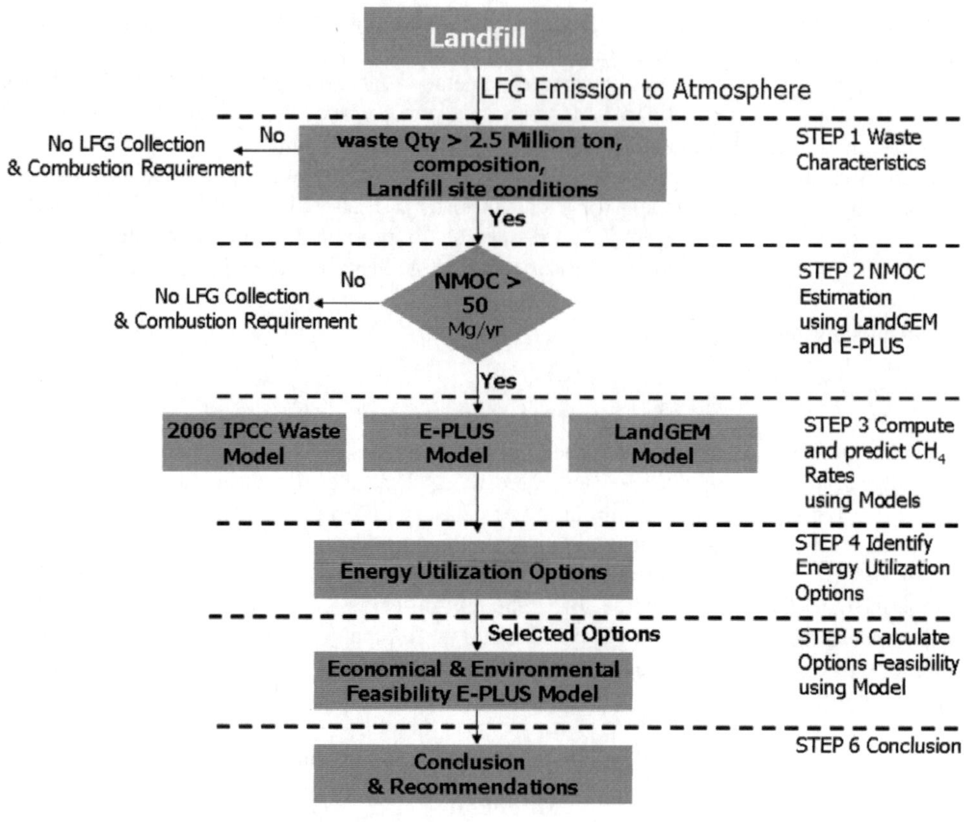

Figure 1. Methodology Algorithm.

2.3 Estimation of Landfill gases

The emission rate for nonmethane organic compounds, NMOCs, will be estimated using the following simulation models: (1) The United States Environmental Protection Agency (USEPA) Landfill Gas Emission Model (LandGEM– version 3.02) (Pelt R., 1998; U. S. Environmental Protection Agency (EPA), 2005). (2) Energy Project Landfill Gas Utilization Software (E-PLUS-version 1.0) (Weitz M., 2007). (3) 2006 IPCC Waste Model (Environment Canada, 2003; Pelt R., 1998; Weitz M., 2007).

2.3.1 LandGEM model. The LandGEM uses the following first-order decomposition rate equation to estimate annual LFG, methane and NMOC generation rates over a time period. The LandGEM model parameters are:

- Methane generation rate (k);
- Potential methane generation capacity (Lo);
- NMOC concentration in landfill gas. It is a function of the types of waste in the landfill and the extent of the reactions that produce various compounds from the anaerobic decomposition of waste. NMOC concentration is measured in units of parts per million by volume (ppmv) and is used by LandGEM only when NMOC emissions are being estimated. An NMOC concentration of 4,000 ppmv (Clean Air Act) as hexane was used;

- Methane content 40% as per landfill data. Although methane production is not affected by the concentration of methane, production of carbon dioxide depends on this value.

The model is described by the following equation:

$$Q_{CH_4} = \sum_{i=1}^{n} \sum_{j=0.1}^{1} kL_o \left[\frac{M_i}{10}\right] e^{-ktij} \quad (1)$$

Q_{CH_4}: annual methane generation in the year of the calculation (m^3/y)
i: 1-year time increment
n: (year of the calculation) - (initial year of waste acceptance)
j: 0.1-year time increment
k: methane generation rate ($year^{-1}$)
L_o: potential methane generation capacity (m^3/Mg)
M_i: mass of waste accepted in the ith year (Mg)
t_{ij}: age of the jth section of waste mass M_i accepted in the ith year (decimal years, e.g., 3.2 years)

2.3.1.1 Estimating k. The values of methane generation rate, k, are given in Table 2.Therefore, the location of El Hammam landfill site is considered an arid area as it receives less than 25 inches (635 mm) of rainfall per year. In line with the USEPA (United State Environmental Protection Agency) recommendation, the k value for our calculation has been selected at 0.02 for arid area.

Table 2. k values.

Default Type	Landfill Type	k value ($year^{-1}$)
CAA (Clean Air Act)	Conventional	0.05
CAA	Arid Area	0.02
Inventory	Conventional	0.04
Inventory	Arid Area	0.02
Inventory	Wet (Bioreactor)	0.7

2.3.1.2 Estimating Lo. The Potential Methane Generation Capacity, L_o, depends only on the type and composition of waste placed in the landfill. The higher the cellulose contents of the waste, the higher the value of L_o and was estimated by the following equation to be 105 m^3 CH_4/ t waste

$$Lo = MCF * DOC * DOC_f * F * (16/12) \quad (2)$$

Lo: Potential methane generation capacity of the waste (ton CH_4 / ton Waste)
MCF : Methane Correction Factor
DOC : Degradable organic carbon in the waste (fraction)
DOC_f : Fraction of organic carbon dissimilated (fraction)
F : Fraction of CH_4 in the landfill gas (fraction)
16/12 : [molecular weight of methane gas to the atomic Weight of carbon]

According to the 2006 IPCC (Intergovernmental Panel on Climate Change) Guidelines for National Greenhouse Gas Inventories, MCF is assumed to be 1 since the landfill is well managed.

Degradable organic fraction (DOC) is based on the composition of the waste. It is estimated from weighted average of the carbon content of various components of the waste stream. Applying the following equation DOC was found to be 0.155

$$\% DOC\ (by\ weight) = 0.4(A) + 0.17(B) + 0.15(C) + 0.30(D) \quad (3)$$

A: is the percent paper and textiles in the waste
B: is the percent garden and park waste, and other non-food organics
C: is the percent food waste
D: is the percent wood and straw waste

2.3.1.3 Estimating CO_2 generation rate. The production of carbon dioxide (Q_{CO2}) is calculated from the production of methane (Q_{CH4}) and the methane percentage (P_{CH4} = 40%) using the following equation:

$$Q_{CO_2} = Q_{CH_4} * 60 / 40 \quad (4)$$

2.3.2 E-PLUS Model. E-PLUS developed two methods to estimate methane, NMOC and LFG generation rates which are WIP-30 and First Order Decay.

2.3.2.1 WIP-30. The WIP-30 method uses the amount of waste in place over the past 30 years to estimate methane emissions.

The equations for small landfills (≤ 907,000 tons of waste in place) and large landfills (>907,000 tons of waste in place) in arid and non-arid regions are outlined in Table 3.

Table 3. WIP-30 Equations for Arid and Non-Arid Regions.

	Arid Region	**Non-Arid Region**
Small Landfills	CH_4 emission = 18.56025 * (5.3253 *10-6 * W)	CH_4 emission = 18.56025 * [6.9492 *10-6 * W)]
Large Landfills	CH_4 emission = 18.56025 * [8.22 + (3.1298 * 10-6 * W)]	CH_4 emission = 18.56025 * [8.22 + (5.0259 * 10-6 * W)]

W = waste in place that is less than 30 years old, (106 tons)
Arid environment = Less than 25 inches of precipitation per year
Non-arid environment = More than 25 inches of precipitation per year

2.3.2.2 First Order Decay. The first order decay equation is the following.

$$Q = Lo * R * (e^{-KC} - e^{-Kt}) \quad (5)$$

Q: Methane generated in current year (m^3/yr).
Lo: Methane generation potential (m^3/Mg of refuse). Default value for arid environment is 2.7228 ft^3 per pound of refuse or 170 m^3 CH_4 /Mg of refuse.
R: Average annual waste acceptance rate during active life (Mg/yr).
K: Methane generation rate constant (yr^{-1}). Default value for arid environment is 0.05 yr^{-1}.
C: time since landfill closure (yr)
t: time since landfill opened (yr)

2.3.3 2006 IPCC Waste Model. The 2006 IPCC waste model provides two options for the estimation of CH_4 emissions from MSW, which can be chosen depending on the available activity data.

2.3.3.1 Waste Composition Model [17,19,24]. CH_4 is generated as a result of degradation of organic material under anaerobic conditions. Part of the CH_4 generated is oxidized in the cover of the SWDS (Solid Waste Disposal Sites), or can be recovered for energy or flaring. The CH_4 actually emitted from the SWDS will hence be smaller than the amount generated. The CH_4 emissions from waste for a single year can be estimated using the following equation.

$$CH_4\ Emissions = [\Sigma x\ CH_4\ generated\ x, T - RT] * (1 - OXT) \quad (6)$$

CH_4 Emissions: CH_4 emitted in year T, Gg
T: Inventory time
x: Waste category or type/material
RT: CH_4 Recovery in year T, Gg
OXT: Oxidation factor in year T, (fraction)

Methane recovery *(R):* CH_4 generated at SWDS can be recovered and combusted in a flare or energy device. The amount of CH_4 which is recovered is expressed as R. The default value for CH_4 recovery is (zero) which is used in our calculations.

Oxidation factor *(OX):* it reflects the amount of CH_4 from SWDS that is oxidized in the soil or other material covering the waste. CH_4 oxidation is by methanotrophic micro-organisms in cover soils and can range from negligible to 100% of internally produced CH_4. The thickness, physical properties and moisture content of cover soils directly affect CH_4 oxidation.

One key input to the model is DOC_m (Degradable Organic material in Gg), calculated on the basis of $DDOC_m$ (Decomposable Degradable Organic material in Gg).

The following two equations are used for calculating $DDOC_m$ accumulated and decomposed at the end of year T. k is the reaction constant.

$$(DDOC_m)a_T = (DDOC_m)d_T + (DDOC_m a_{T-1} * e^{-k}) \tag{7}$$

Where

DDOCmaT	DDOCm accumulated in the SWDS at the end of year T, Gg
T	Inventory time
DDOCmdT	DDOCm deposited into the SWDS in year T, Gg
DDOCmaT-1	DDOCm accumulated in the SWDS at the end of year (T-1), Gg
k	Reaction constant, k = ln(2) / t1/2, (y-1)
t1/2	Half-life time (y)

$$(DDOC_m)d_T = (DDOC_m)a_{T-1} * (1 - e^{-k}) \tag{8}$$

Generated Methane can then be calculated from the following equation, where F is the fraction of Methane in the generated gas:

$$CH_4\ generated_T = DDOC_m\ d_T * F * 16/12 \tag{9}$$

The parameters used as input to the model are shown in Table 4.

2.3.3.2 Bulk Waste Model. The only differences in bulk option calculations are the following:

- DOC value for bulk MSW = 0.12 Gg
- Methane generation rate constant (k) = 0.05

Input values are shown in Table 5.

2.4 Comparison between the different models

Table 6 presents a comparison between the different models used.

3. Results and discussion

3.1 Simulation Results for NMOC, LFG and methane generation

The average NMOC emitted from El Hammam landfill is 367 Mg/yr and 740 Mg/yr as calculated by both LandGEM and E-PLUS model, respectively. The maximum value for NMOC emitted during landfill lifetime is 577 Mg/yr and 1,240 Mg/yr as predicted by both LandGEM and E-PLUS models respectively. Figure 2 illustrates the NMOC annual generation rates as predicted by simulation models.

The emission rates for methane and landfill gas have been estimated using the three simulation models and the results are presented in Figure 3. The results show that the total CH_4 flow during 2003-2045 ranges between 1.018 x 10^{10} ft^3/yr based on the 2006 IPCC bulk option and 2.156 x 10^{10} ft^3/yr based on the E-PLUS WIP-30 method. Maximum methane generation rate during 2003-2045 ranges between 4.129 x 10^8 ft^3/yr based on the E-PLUS model – WIP30 calculation method and 7 x 10^8 ft^3/yr based on the 2006 IPCC model – Bulk Option. The average annual methane generation rate during 2003-2045 ranges between 2.367 x 10^8 ft^3/yr based on the 2006 IPCC bulk option and 5.014 x 10^8 ft^3/yr based on the E-PLUS WIP-30 method. Figure 4 represents the results of such simulations.

Table 4. The 2006 IPCC Waste Model Input Parameters, Waste Composition Option.

ntry fx El Hammam - ONYX, Alexandria, Egypt

Parameters

Country El Hammam - ONYX, Alexandria, Egypt
Region Africa: Northern

Please enter parameters in the yellow cells. If no national data are available, copy the IPCC default value.
Help on parameter selection can be found in the 2006 IPCC guidelines

B	C	D	E	F
hab El-Sersy	**IPCC default value**		**Country-specific parameters**	
			Value	**Reference and remarks**
Starting year		2003	2003	
DOC (Degradable organic carbon)	Waste by composition			
(weight fraction, wet basis)	**Range**	**Default**		
Food waste	0.08-0.20	0.15	0.15	
Garden	0.18-0.22	0.2	0	
Paper	0.36-0.45	0.4	0.4	
Wood and straw	0.39-0.46	0.43	0.43	
Textiles	0.20-0.40	0.24	0.24	
Disposable nappies	0.18-0.32	0.24	0	
Sewage sludge	0.04-0.05	0.05	0	
Industrial waste	0-0.54	0.15	0	
DOCf (fraction of DOC dissimilated)		0.5	0.5	
Methane generation rate constant (k)	Dry temperate			
(years^{-1})	**Range**	**Default**		
Food waste	0.05-0.08	0.06	0.06	
Garden	0.04-0.06	0.05	0	
Paper	0.03-0.05	0.04	0.04	
Wood and straw	0.01-0.03	0.02	0.02	
Textiles	0.03-0.05	0.04	0.04	
Disposable nappies	0.04-0.06	0.05	0	
Sewage sludge	0.05-0.08	0.06	0	
Industrial waste	0.04-0.06	0.05	0	
Delay time (months)		6	6	
Fraction of methane (F) in developed gas		0.5	0.37	
Conversion factor, C to CH_4		1.33	1.33	
Oxidation factor (OX)		0	0.1	

Instructions / **Parameters** / MCF / Activity / Amnt Deposited / Recovery O

Table 5. The 2006 IPCC Waste Model Input Parameters; Waste Bulk Option.

ountry ▾ fx El Hammam - Bulk Option, ONYX, Alexandria, Egypt

B	C	D	E	F
Parameters	**Country**	**El Hammam - Bulk Option, ONYX, Alexandria, Egypt**		
	Region	Africa: Northern		
Please enter parameters in the yellow cells. If no national data are available, copy the IPCC default value.				
Help on parameter selection can be found in the 2006 IPCC guidelines				
Ihab El-Sersy	**IPCC default value**		**Country-specific parameters**	
			Value	**Reference and remarks**
Starting year		2003	2003	
DOC (Degradable organic carbon)	Bulk waste data only			
(weight fraction, wet basis)	**Range**	**Default**		
Bulk MSW	0.12-0.28	0.16	0.12	
Industrial waste	0-0.54	0.15	0	
Sewage sludge	0.04-0.05	0.05	0	
			0	
			0	
			0	
			0	
			0	
DOCf (fraction of DOC dissimilated)		0.5	0.5	
Methane generation rate constant (k)	Dry temperate			
(years⁻¹)	**Range**	**Default**		
Bulk MSW	0.04-0.06	0.05	0.05	
Industrial waste	0.04-0.06	0.05	0	
Sewage sludge	0.05-0.08	0.06	0	
			0	
			0	
			0	
			0	
			0	
Delay time (months)		6	6	
Fraction of methane (F) in developed gas		0.5	0.37	
Conversion factor, C to CH_4		1.33	1.33	
Oxidation factor (OX)		0	0.1	

Parameters / MCF / Activity / Amnt_Deposited / Recovery_OX / Results / HWP

Table 6. Comparison between the different models.

Parameters/ Assumptions	LandGEM	E-PLUS		2006 IPCC	
		FOD	WIP-30	Compositio n	Bulk
Rate equation	1st Order	1st Order	Empirica l	1st Order	1st Order
Landfill Closure Year	2026	2026	2026	2026	2026
Model Inputs	K as bulk, Waste Qty per year	K as bulk, Waste Qty per year		DOC and K for waste comp. Waste Qty per capita	DOC and K for bulk waste. Waste Qty per capita
Model Outputs	LFG, CH_4, CO_2 , NMOC	LFG, CH_4, NMOC. Environmental Benefits Analysis. Economical Analysis		CH_4 ONLY total and for each waste component	CH_4 ONLY total
Comments	Simple Approach	Simple Approach		Complex	
	Integrates emissions over a 0.1 year time increment.	Design configuration and evaluation of energy utilization options.		Compute for several sites MSW, Industrial waste and sewage sludge.	

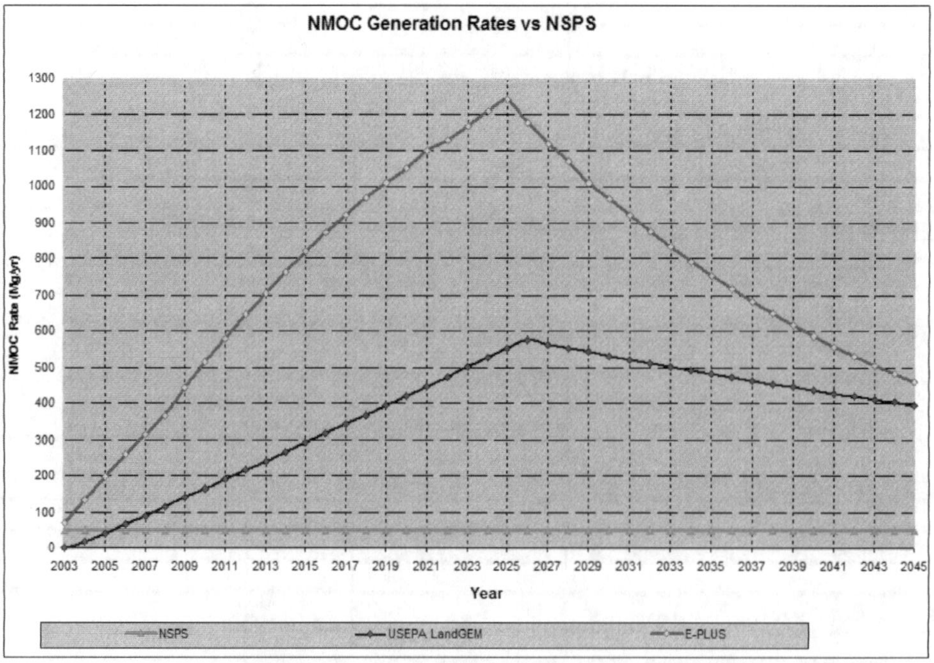

Figure 2. NMOC generation rate using the different models.

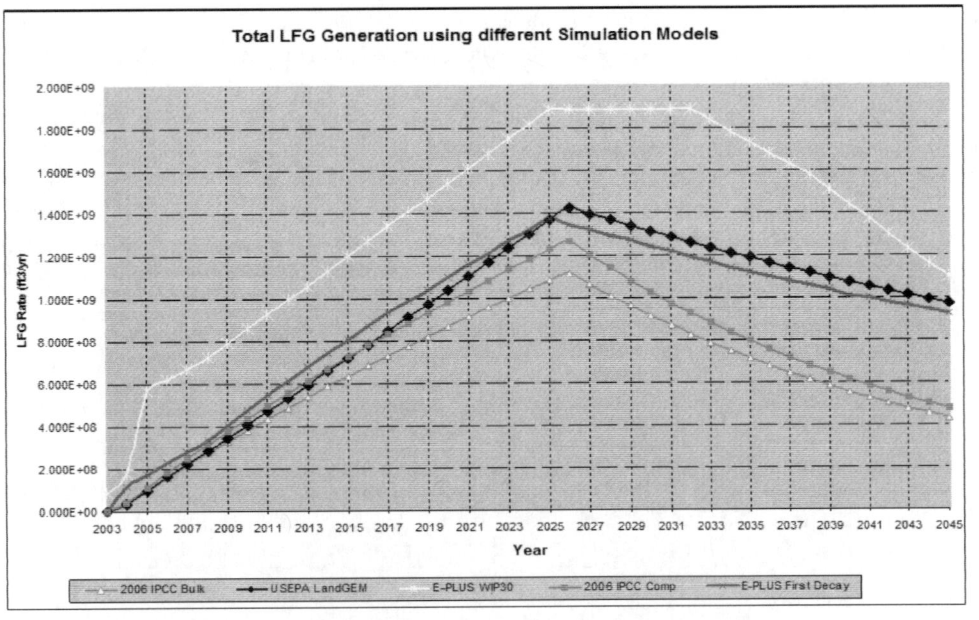

Figure 3. LFG generation rate using the different models.

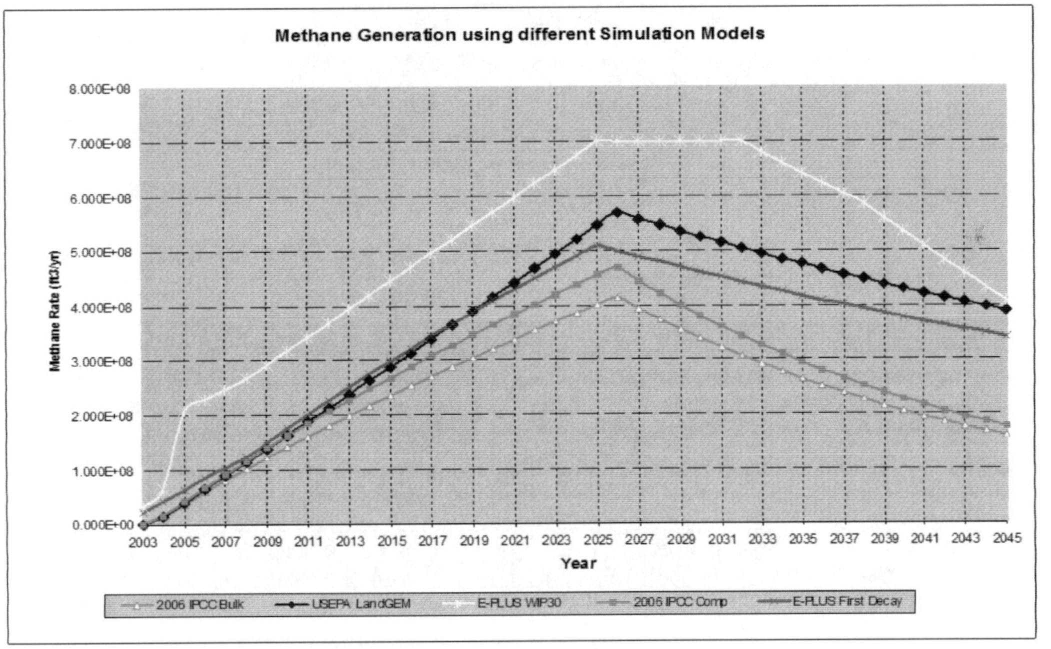

Figure 4. Methane generation rate simulation results.

3.2 Model validation

The current collection and flaring system in El Hammam landfill site implemented in August 2006 is assumed to have a collection efficiency of 70% [13] with a maximum design inlet LFG flow rate to the flaring system equivalent to 2500 m^3/h. This means that approximately 1.104 x 10^9 ft^3/y is generated at the landfill site assuming the same collection efficiency in order to meet the existing inlet flare design flow rate. We can conclude from the results presented in Graph 4 that all annual LFG generation rate estimated by simulation models are in coherence with the existing flare design flow rate at 70% collection efficiency except for the over estimated value from the E-PLUS model – WIP30 method (1.323 x 109 ft^3/y).

Furthermore, the 2006 IPCC model – Bulk Option seems to be under estimating the LFG generation rate. These two methods are also presenting overestimating and underestimating values as well for methane generation rate as illustrated in Figure 5. Consequently, these two methods are suggested to be excluded from further evaluation in order to ensure conservativeness. 2006 IPCC Waste Model. E-PLUS First-Order Decay Method was found to be the more conservative method in assuming the maximum methane generation rate per year over the studied period from 2003-2045.

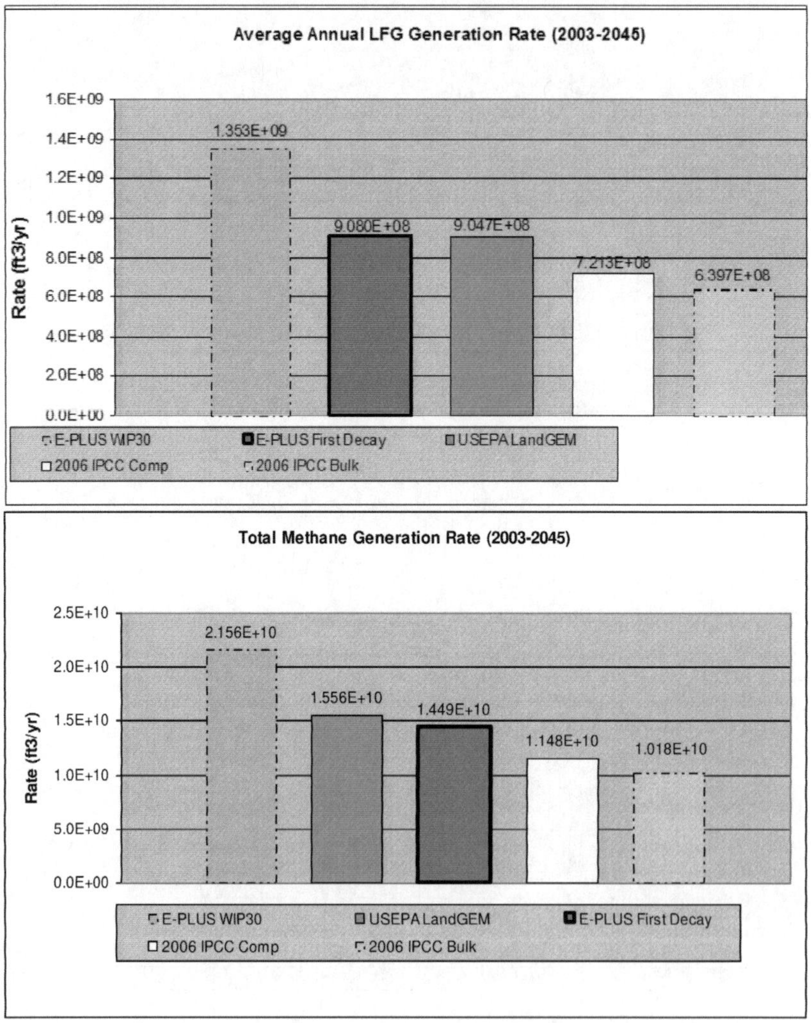

Figure 5. top: LFG generation rate and bottom Methane generation rate.

4. Conclusion

Methane generation rates from Alexandria landfill site in El Hammam have been estimated using three different simulation models namely: LandGEM, E-PLUS, and the 2006 IPCC Waste Model. E-PLUS First-Order Decay Method was found to be the more conservative method in assuming the maximum methane generation rate per year over the studied period from 2003-2045. Therefore, this conservative method is suggested for evaluating different energy utilization options.

References

[1] Abbasi T and Abbasi S A 2010 Biomass energy and the environmental impacts associated with its production and utilization *Renew. Sustain. Energy Rev.* **14** 919–37

[2] Bilen K, Ozyurt O, Bakırcı K, Karslı S, Erdogan S, Yılmaz M and Comaklı O 2008 Energy production , consumption , and environmental pollution for sustainable development : A case study in Turkey *Renew. Sustain. Energy Rev.* **12** 1529–61

[3] Bilgen S, Keleş S, Kaygusuz A, Sarı A and Kaygusuz K 2008 Global warming and renewable energy sources for sustainable development: A case study in Turkey *Renew. Sustain. Energy Rev.* **12** 372–96

[4] Naqvi S M K and Sejian V 2011 Global climate change: role of livestock *Asian J. Agric. Sci.* **3** 19–25

[5] Russell R 2007 The Greenhouse Effect & Greenhouse Gases *Wind. to Universe Univ. Corp. Atmos. Res.*

[6] Yusuf R O, Noor Z Z, Abba A H, Hassan M A A and Din M F M 2012 Methane emission by sectors: a comprehensive review of emission sources and mitigation methods *Renew. Sustain. Energy Rev.* **16** 5059–70

[7] Yüksel I 2008 Global warming and renewable energy sources for sustainable development in Turkey *Renew. Energy* **33** 802–12

[8] Hansen J, Sato M, Kharecha P, Russell G, Lea D W and Siddall M 2007 Climate change and trace gases *Philos. Trans. R. Soc. London A Math. Phys. Eng. Sci.* **365** 1925–54

[9] Lakeman J A 1996 *Climate change 1995: the science of climate change: contribution of working group I to the second assessment report of the Intergovernmental Panel on Climate Change* vol 2 (Cambridge University Press)

[10] Johannessen L M 1999 *Guidance note on recuperation of landfill gas from municipal solid waste landfills* (World Bank, Urban Development Division, Urban Waste Management Thematic Group)

[11] Zhu H, Letzel M O, Reiser M, Kranert M, Bächlin W and Flassak T 2013 A new approach to estimation of methane emission rates from landfills *Waste Manag.* **33** 2713–9

[12] Metz B 2007 *Climate change 2007: mitigation: contribution of working group III to the fourth assessment report of the intergovernmental panel on climate change* (Intergovernmental Panel on Climate Change)

[13] (PDD) C P D D 2006*Onyx Alexandria Landfill Gas Capture and Flaring Project. Borg El Arab and El Hammam Landfills . Alexandria, Egypt*

[14] Alcamo J, Bouwman A, Edmonds J, Grubler A, Morita T and Sugandhy A 1995 An evaluation of the IPCC IS92 emission scenarios *Clim. Chang. 1994*

[15] Huber-Humer M 2004 International research into landfill gas emissions and mitigation strategies--IWWG working group" CLEAR". *Waste Manag. (New York, NY)* **24** 425–7

[16] Alexander, Amy, C. E. Burklin and A S 2005 *Landfill gas emissions model (LandGEM) version 3.02 user's guide*

[17] Weitz, Melissa, Jeffrey B. Coburn and E S 2008 Estimating national landfill methane emissions: an application of the 2006 Intergovernmental Panel on Climate Change waste model in Panama. *J. Air Waste Manage. Assoc.* **58** 636–40

[18] Xydis G, Nanaki E and Koroneos C 2013 Exergy analysis of biogas production from a municipal solid waste landfill *Sustain. Energy Technol. Assessments* **4** 20–8

[19] Olsen K, Wellisch M, Boileau P, Blain D, Ha C, Henderson L, Liang C, McCarthy J and McKibbon S 2003 Canada's greenhouse gas inventory 1990-2001 *Environ. Canada Greenh. Gas Div. Ottawa,* 205

[20] Galle B O, Samuelsson J, Svensson B H and Borjesson G 2001 Measurements of methane emissions from landfills using a time correlation tracer method based on FTIR absorption spectroscopy *Environ. Sci. Technol.* **35** 21–5

[21] Di Bella G, Di Trapani D and Viviani G 2011 Evaluation of methane emissions from Palermo municipal landfill: Comparison between field measurements and models *Waste Manag.* **31** 1820–6

[22] Amini H R, Reinhart D R and Mackie K R 2012 Determination of first-order landfill gas modeling parameters and uncertainties *Waste Manag.* **32** 305–16

[23] Ishii K and Furuichi T 2013 Estimation of methane emission rate changes using age-defined waste in a landfill site *Waste Manag.* **33** 1861–9

[24] Pelt R 1998 *User's Manual Landfill Gas Emissions Model, Version 2.0* (US Environmental Protection Agency, Air Pollution Prevention and Control Division)

[25] Gaur A, Park J-W, Maken S, Song H-J and Park J-J 2010 Landfill gas (LFG) processing via adsorption and alkanolamine absorption *Fuel Process. Technol.* **91** 635–40

[26] I.M. E S 2008*Maximum Methane Yield in Landfill Gas to Enhance Economic Feasibility for Energy Utilization* (Cairo University, Egypt)

[27] Morris J W F and Barlaz M A 2011 A performance-based system for the long-term management of municipal waste landfills *Waste Manag.* **31** 649–62

[28] Anon The Bioreactor Landfill – the Next Generation of Landfill Management

Synthesis, structural elucidation and antimicrobial activities of 5-(3-nitrophenyllazo)-6-aminouracil and its complexes with some transition metal ions

Z M Zaki[1], S M Abbas[1], H A Dessoukii [2], H S Awes[1], Rehab Mahmoud[3*]

[1] Inorganic Chemistry Lab. Department of Chemistry, Faculty of Science, Beni-Suef University, Beni-Suef, Egypt.
[2] chemistry department, faculty of science, Benha university, Eygpt
[3] Materials Science Lab, Chemistry Department, Faculty of Science, Beni-Suef University, Beni Suef, Egypt

Abstract. In this study, 5-(3-nitrophenyllazo)-6-aminouracil was prepared by linking 3 -nitro aniline to 6-aminouracil. The (1:1) complexes of this ligand with Fe^{+3} , Co^{+2},Ni^{+2}, Cu^{+2} and Zn^{+2} were prepared and characterized by elemental analysis for C, H and N, thermal analyses (TGA and DTA) , FT-IR, ^{1}H NMR. The molecular weight of the ligand was determined by mass spectral measurement and its fragmentation pattern was reported. The metal content of the complexes was determined by EDTA complexometric titrations and the coordination number of metal in complexes was confirmed by magnetic susceptibility measurements. Antimicrobial activities for ligand and its complexes were studied. The IR and ^{1}H NMR studies confirmed that 5-(3-nitrophenyllazo)-6-minouracil behaves as monobasic bi dentate ligand bonding to the metal ion through both NH_2 group and N=N groups forming six membered ring and The magnetic moments measurements indicated octahedral geometry for Fe^{+3} complex, tetrahedron structure for Co^{+2}, Ni^{+2}, Cu^{+2} complexes and square planar geometry for Zn^{+2} complex. Antimicrobial activity measurements indicated that Co^{+2} and Ni^{+2} complexes are biologically active towards bacteria gram (-ve), gram (+ ve) and fungi, while Zn^{+2} complex showed high resistance towards gram (+ ve) bacteria and medium resistance towards gram (-ve).

Keywords: Uracils, synthesis, Uracil azo dyes, azo dye complexes, transition metal complexes, mass spectra, IR, NMR, Thermal analysis, biological activity

1. Introduction

The compounds containing the functional group R-N=N-R~ are known as azo compounds. Both R and R` may be either aryl or alkyl radicals forming different colors due to mobility of labile proton across the molecule through a conjugated system where the proton remains in association with intra molecular hydrogen bonding between donor –acceptor atoms through aromatic moieties [1–3] and have many application dependent on the photo physical properties of azo hydrazone tautomerism that used in dyes of surfactant micelles and textile fibers [4,5], photographic systems [6], dyeing protein [7,8] and bleaching [9].The diazonium salts are stabilized by resonance and react as an electrophile with an electron rich coupling component like phenols and anilines through an electrophilic aromatic substitution mechanism. Some heterocyclic azo compounds are used as ligands to generate a special category of metal azo complexes which are exploited enormously in the manufacture of colorimetric sensors [10]. 6-amino uracil has different reactive sites, therefore has different binding modes so, it can coordinate through one of pyrimidine ring nitrogen, the C=O or NH_2 [11,12]. It is used as a starting material for synthesis of heterocyclic frameworks of biological significance such as pyrido-pyrrolo and pyrimidopyrimidines [13–15].

Synthesis and study of metal complexes of heterocyclic compounds have wide range of applications [16] especially the compounds containing nitrogen and sulfur as donor atoms whose anticancer and antiviral activities [17] The Antimicrobial activity is one of the most important activities for exploring novel antimicrobial agents that control the resistant pathogens due to overuse of antibiotics. An example for gram negative bacteria is *Escherichia coli*, which is one member of the normal flora in the intestine. Under stress *E.coli* a bacterium become pathogenic and causes severe intestinal, urinary, biliary and abdominal infections. Other than *E.coli* pathogenicity t produces potent and severs toxins very harmful and dangerous for the small intestinal wall [18,19].

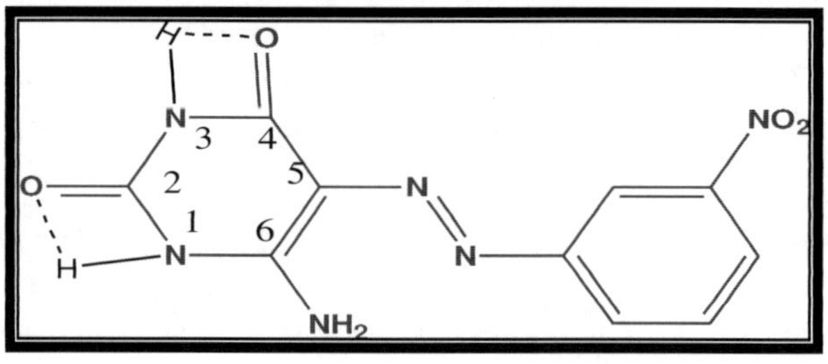

Scheme 1. 5-(3-nitrophenyllazo)-6-aminouracil

On the other hand, *Staphylococcus aureus* is a gram-positive bacterium which is normal inhabitant in the respiratory tract and on the skin. *S. aureus* at skin infection become pathogenic and causes sinusitis, and food poisoning besides producing virulence factors such as potent protein toxins [20].

In this study, metal complexes of 5-(3-nitrophenyllazo)-6-aminouracil with Fe(III), Co(II), Ni(II) and Cu(II) were prepared and subjected to many tools including elemental analysis for C, H and N, thermal analyses (TGA and DTA), FT-IR, ^1H NMR. to confirm their structure. The study excluded chelation either through N(3) and C(2)=O, or through N(3) and C(4)=O due to hydrogen bonding [21–25] and revealed that the ligand coordinates to the metal cations in bidentate fashion by coordinating to amino group (deprotonated N6) of uracil ring and azo group. The ligand and complexes were subjected to antimicrobial activity investigation against pathogenic *E.coli*, *Klebsiella pnemonia and Pseudomonas aeruginosa* as Gram-negative bacteria while *S. aureus* and *S.mutans* as Gram-positive bacteria but *candida albicans* for antifungal evaluation for efficacy enhancement and decreasing the resistance.

2. Materials and method

2.1. Materials and reagents

All chemicals used are of the highest purity grade and the solvents used were spectroscopic grade. 5-(3-Nitro phenylazo)-6-aminouracil was prepared by using 3-nitro aniline, Sodium nitrite and 6-

aminouracil that were purchased from MERCK-Schuchardt, SDFCL-INDIA and EUROMEDEX FRANCE and hydrochloric acid (HCl) were used from Carlo Erba reagent NaOH Piochem for laboratory chemicals, EGYPT, While Metal Chloride salts like : $FeCl_3$ $6H_2O$, $CoCl_2.6H_2O$, $NiCl_2.6H_2O$, $CuCl_{2.6}H2O$ and $ZnCl_2.6H_2O$.were purchased from Aldrish chemical company .

2.2 Instrumentation

The infrared spectral data of the ligand and their complexes were taken in Potassium bromide disc using Perkin Elmer spectrophotometer serial number 1341 covering frequency range 400-4000 cm^{-1}. Proton nuclear magnetic resonance (^{1}HNMR) spectra of the ligand and its complexes were determined with a Bruker FT-NMR instrument at 400.1324 MHz in DMSO-d6 with TMS as internal standard spectrometer at NMR laboratory at faculty of pharmacy at Beni-Suef University. Elemental analyses for C, H and N were recorded by Vario EL III Germany Laboratory, Mass spectra were determined on GC MS-Qp1000EX Shimadzu and thermal analysis(TG-DTG)(20-1100^{0}C) was performed with heating rate of 10^{0}C min^{-1} and recorded on Shimadzu thermogravimetric analyzer (TGA-50H) and (DTG-60H) at Microanalytical unit of Cairo university. Magnetic susceptibility measurements were carried out on powder samples and were evaluated at room temperature using The GOUY Method that carried out on Johnson Matthey magnetic susceptibility balance at central laboratory of chemistry department while the Antimicrobial activities of ligand and its complexes were examined_at biochemistry central laboratory of chemistry department at Cairo University.

2.3. Preparation of 5-(3-Nitrophenylazo)-6-aminouracil

5-(3-Nitro phenylazo)-6-aminouracil was prepared by diazotization of 3-nitro aniline 0.04 mole (5.5 g) dissolved in 6 mL of concentrated HCl and 30 mL of water (26). The solution was cooled down in a salt/ice bath, and then a cold solution of $NaNO_2$ 0.04 mole, (2.76 g) in 3 mL of water was added in a dropwise manner with constant stirring. The resulting diazonium salt was cooled in a salt/ice bath and then, 0.04 mole (5 g) of 6-Aminouracil dissolved in 25(ml) (2N) NaOH solution and cooled in salt/ice bath was added to the cold diazonium solution in a dropwise manner with constant stirring. The solution was stirred at 0-5° C for 1 h. and the pH of the reaction mixture was maintained at 4-6 by the simultaneous addition of 2ml of glacial acetic acid. The resulting precipitate was filtered off, washed with cold water and ethanol and dried. Recrystallization from mixture containing (150 ml ethanol, 150 bidistilled water and 2ml of 25% aqueous ammonia solution [26].

The characterization data are listed in Table 1 which shows a good agreement between calculated and found data for the elemental analysis of the prepared compound

2.4. Preparation of solid 5-(3-Nitro phenylazo)-6-aminouracil complexes

The complexes were prepared by taking 0.01mole of the metal chloride salt in 5ml of mixture (1:1) aqueous ethanol was added to 0.01 mole of solution of 5-(3-Nitro phenylazo)-6-aminouracil that dissolved in hot mixture containing (150 ml ethanol, 150 bidistilled water and 2ml of 25% aqueous ammonia solution) the mixture was stirred for 30 minute at 60^{0}C. The precipitated complex was filtered off, washed several times by ethanol (95%) and dried in the oven. The physical properties of all complexes were listed in table 1

2.5. Metal analysis

Metal content in the complexes was determined by digesting accurate weight (0.03 gm.) of complex in concentrated aqua regia solution ($HCl+HNO_3$) then titrimetric with standard EDTA (0.01M) using suitable indicator [27,28].

Table 1. Analytical data and physical properties of the5(3-nitro phenyllazo)6- amino uracil ligand and its prepared complexes

	Ligand	Fe^{+3}	Co^{+2}	Ni^{+2}	Cu^{+2}	Zn^{+2}
M. wt.	276	400.84	368.93	404.69	409.54	375.41
M. F	$C_{10}H_8N_6O_4$	$C_{10}H_{13}N_6O_8$ Fe	$C_{10}H_{10}N_6$ O_6Co	$C_{10}H_{14}N_6$ O_8Ni	$C_{10}H_{14}N_6$ O_8Cu	$C_{10}H_{10}N_6$ O_6 Zn
Color	Canary yellow	Brownish	Dark brownish	Brownish yellow	Brown	Yellow
% C_{Theo} % C_{exp}	43.47 40.51	29.93 29.63	32.52 31.90	29.56 29.41	29.30 29.12	31.96% 31.48%
$\%H_{Theo}$ $\%H_{exp}$	2.89 2.43	1.7 1.6	2.71 2.53	3.45 3.33	3.41 3.33	2.66% 2.24%
$\%N_{Theo}$ N_{exp} %	30.43 28.82	20.95 21.22	22.72 22.21	20.75 20.83	20.51 20.44	22.37 21.85
$\%M_{Theo}$ $\%M_{exp}$		13.93 12.51	15.97 16.05	14.50 14.41	15.51 15.90	17.42% 16.96%
Purposed structure	H_2L	$Fe(H_2L)(OH)_2(H_2O)_2$	$CoH_2L(OH)(H_2O)$	$NiH_2L(OH)(H_2O).2H_2O$	$Cu(H_2L)(OH)(H_2O).2H_2O$	$ZnH_2L(OH)(H_2O)$

2.6 Isolation of bacterial and fungal pathogens

Bacterial strains were grown on its selective media as *E.coli* on MacConkey agar media (Oxoid; CM 0115) and EMB; Oxoid; CM 69, plates whereas, the other strains (*Staphylococcus* and *Streptococcus*) were determined on agar plates and were grown on their nutrient media. The plates were incubated at 37°C for 48 hours and the bacterial growth was observed for all of the isolates [29].

2.7 Biochemical and serological identification of bacterial pathogens

Standard Kits (Biomerieux, Marcy L'etoil, France) API had been used for both physiological and biochemical identification of each bacterial isolates of both gram positive and negative bacteria.

2.8 Minimal inhibitory concentration (MIC) measurement

For each strain, three to five isolated colonies were selected from the fresh agar plate and were transferred into a tube containing 3-4 ml of sterile broth medium. The bacterial suspension was mixed well and incubated at 35-37°C for 2-6 h. The turbidity of the bacterial suspension should be equal to or greater than the turbidity of a McFarland Standard 0.5. After that, 1mg of the tested compound (antimicrobial agent) was dissolved in 1 ml DMSO and two-fold serial dilution was done using broth medium. A fixed volume of the prepared bacterial inoculum was added to each tube and incubated for at 37°C 16-20 h. The MIC is defined as the lowest concentration of the antimicrobial agent that inhibits visible growth of the tested isolate as observed with the unaided eye [30].

2.9 Antimicrobial assay

The antimicrobial activity of synthesized compounds was determined using agar well diffusion method[75]. All the compounds were tested *in vitro* for their antibacterial activity against *staphylococcus aureus* and *Streptococcus mutans*(Gram positive bacteria), *Escherichia coli*, *Pseudomonas aeruginosa* and *klebsiella* (Gram negative bacteria) using nutrient agar medium. Ampicillin and Gentamicin were used as standard drugs for Gram positive and Gram negative respectively. DMSO was used as solvent control. The compounds were tested at a concentration of 15 mg/ml against both bacterial and fungal strains [31].

2.10 Method of testing

The sterilized media was poured onto the sterilized Petri dishes (20-25 ml, each petri dish) and allowed to solidify at room temperature. Microbial suspension was prepared in sterilized saline equivalent to McFarland 0.5 standard solution (1.5x 10^5 CFU mL^{-1}) and its turbidity was adjusted to OD= 0.13 using spectrophotometer at 625 nm. Optimally, within 15 minutes after adjusting the turbidity of the inoculum suspension, a sterile cotton swab was dipped into the adjusted suspension and was flooded on the dried agar surface then allowed to dry for 15 minutes with lid in place. Wells of 6 mm diameter was made in the solidified media with the help of sterile borer. 100 μL of the solution of the tested compound was added to each well with the help of micropipette. The plates were incubated at 37°C for 24 hrs in case of antibacterial activity. This experiment was carried out in triplicate and zones of inhibition were measured in mm [31].

3. Results and discussion

3.1. Mass spectra

The mass spectrum of the prepared of 5-(3-nitrophenyazo)-6-aminouracil ligand is shown in Scheme 2 and table 2. The mass spectral pattern of 5-(3-nitro phenyazo)-6-aminouracil showed an intense molecular ion peak at m/z 276 of relative intensity (100%),corresponding to the molecular weight of the compound with the formula $C_{10}H_8N_6O_4$ this intensity is a proof of its quite elevated stability. The molecular ion undergoes cleavage of C-N bond adjacent to the azo group giving two fragments with two ion peaks at m/z 154 and m/z 122 [32] corresponding to $[C_4H_4N_5O_2]^+$ and $[O_2N\text{-}C_6H_4]^+$, respectively. The fragment $[O_2N\text{-}C_6H_4]^+$ loses nitric oxide molecule producing the fragment $[C_6H_4O]^+$ at m/z 92 and this loss is characteristic for an aromatic nitro group and this peak is usually weak but it is usually present [33] . The latter ion fragments undergoes further loss of carbon monoxide molecule producing $[C_5H_4]^+$at m/z 63 [33] . The ionic fragment $[C_4H_4N_5O_2]^+$ loses N_2 molecule to give the fragment $[C_4H_4N_3O_2]^+$ with m/z 126 which loses a stepwise two CO molecules producing the fragments $[C_3H_4N_3O]^+$ at m/z 98 and $[C_2H_4N_3]^+$ at m/z 70. The peaks observed at m/z 83 and at m/z 55 can be attributed to the loss of iso cyanic acid [HNCO] from the aromatic cycle of the uracil ring $[C_4H_4N_3O_2]^+$ making a retro Diels-Alder mechanism [34] producing the ionic fragment $[C_3H_3N_2O]^+$ which losses CO giving the ionic fragment $[C_2H_3N_2]^+$ m/z 55 [26,35].

3.2. Infrared spectroscopy

The infrared spectral data of the5-(3-nitrophenylazo)-6-aminouracil ligand and its prepared complexes are given in figure 1 and table 3. The main purpose of the IR studies is to assign the vibrational frequencies of the fundamental groups (such as NH_2, NH, C=O, C-N and OH) of the ligand and by comparison to the infrared spectra of complexes to determine the binding sites. The infrared spectral data of the 5-(3-nitrophenylazo)-6-aminouracil ligand showed a broad band in the range 3442-3106 cm^{-1} which was assigned to OH, NH and NH_2 groups [36–38], so it can be suggested that this compound exists in the enol form [36] as shown in scheme 3. The broadness of the band may be suggested that NH group is associated in intramolecular hydrogen bond with the adjacent keto groups.(scheme 3). The band observed at 2800-2813 cm^{-1} in both ligand and complexes is attributed to the aromatic CH stretching frequency. The symmetric deformation mode (δ) of NH_2 and plane bending of N_1H and N_3H groups gives rise to its characteristic frequencies at 1511-1532 cm^{-1} for the synthesized ligand and its complexes [39–41].

Table 2. Relative intensity of molecular ion and major fragments of 5-(3-nitrophenylazo)-6-aminouracil ligand

M. Wt. of fragment	Relative intensity of molecular ion >5%	Assignment
276	100.00	$C_{10}H_8N_6O_4$
154	25.06	$C_4H_4N_5O_2^+$
126	11.64	$C_4H_4N_3O_2^+$
122	20.76	$C_6H_4NO_2^+$
98	78.54	$C_3H_4N_3O^+$
92	19.54	$C_6H_4O^+$
83	18.77	$C_3H_3N_2O^+$
76	18.60	$C_6H_4^+$

Table 3 (a). Assignment of IR bands (Cm^{-1}) of 5- (3-nitro phenyl azo)6- amino uracil ligand and its prepared complexes

Ligand	Fe^{+3}	CO^{+2}	Ni^{+2}	Cu^{+2}	Zn^{+2}	Assign ment
--------- 3442(mb)	3594(w) 3342-3000	3686- 3000(mb)	3689(w) 3664(mb) 3282	3610(w) 3306(m)	3527(w) 3289	¥NH2, ¥OH of H2O and OH
3161(m)	3175(mb)	3258(mb)	3198	3123(m)	3098(m)	¥NH amide
3100(w)	3085(mb)	2977(w)	2980(s)	2977(m)	3007(w)	¥CH aromatic
2813(w)	2806(w)	2886(w)	2903(m)	2800(w)	2813(w)	Strong H bond complex
1744(s)	1737(w)	1725(m)	1702(w)	1727(w)	1737(m)	¥C_2=O amide

Table 3 (b). Assignment of IR bands (Cm^{-1}) of 5- (3-nitro phenyl azo)6- amino uracil ligand and its prepared complexes

1633(s)	1625(m)	1627(m)	1605(m)	1621(m)	1600(m)	¥ C_4=O +¥ C=C
1532(m)	1523(m)	1520(s)	1517(m)	1511(s)	1518(m)	δ NH_2+ δ N_1H a symmetricNO_2
1449(w)	1437 (w)	1396(s)	1389(m)	1409(m)	--------	N=N, δN_3H,
1350(w)	1346(m)	1342(s)	1346(m)	1347(m)	1344(m)	symmetricNO_2
1257(m)	1255(W),	1247(w)	1238(m)	1216(m)	1264(w)	¥ ring stretching C-N
1093(m)	1090(w)	1082(m)	1065(m)	1072(m)	1081(w)	P NH_2C-O,C-N
892(w)	888(w)	876(s)	886(m)	888(m)	889(w)	¥ ring stretching
811 Bending 740 675	807(m) 733(w) 666(w)	801(m) 730(m) 665(s)	808(w) 736(w) 670(w)	814(w) 740(m) 681(m)	803(s) 735(w) 665(m)	¥ ring stretching due to1,3 di substitution +WNH_2
631(m)	-----	-------	-----	-----	--------	¥ N-C=O
----------	514(m)	502(m)	586(w)	576(s)	520(m)	T NH_2, Metal-Nitrogen
---------	425(w)	426(m)	495(m)	434(m)	421(w),459(w)	Metal- Oxygen

The two sharp strong intense peaks observed at 1744 and 1633 cm^{-1} are attributed to the carbonyl groups at C_2=O and C_4=O positions of the uracil ring respectively [42]. The bending deformation vibration band of the C-NH_2 is observed at 1532 cm^{-1} and the stretching vibration band of the azo group is assigned to be at 1449 cm^{-1} [43–47]. The band observed at 1257 cm^{-1} is due to stretching vibration of C-N in the ligand. The presence of 3 bands at 811,740 and 675 cm^{-1} in the ligand spectrum are due to C-NO_2 at the meta position of benzene ring. The infrared spectra of the complexes showed a strong broad band extended from 3689 - 3000 cm^{-1} corresponding to the coordinated H_2O involved in the complexation to satisfy the coordination number and OH group involved in the complexes to satisfy the oxidation number of metal cation. This broad band involves also the stretching bands of NH and NH_2 and phenolic CH of uracil ring.

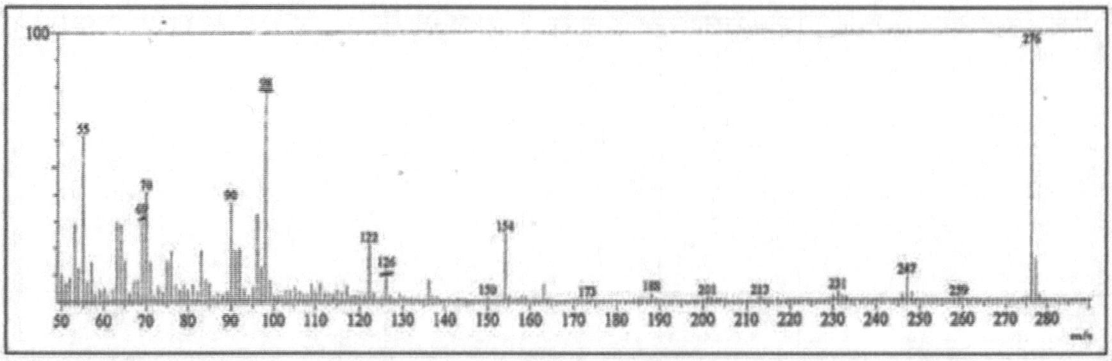

Figure 1. Mass spectrum of 5-(3-nitro-phenylazo)-6-aminouracil

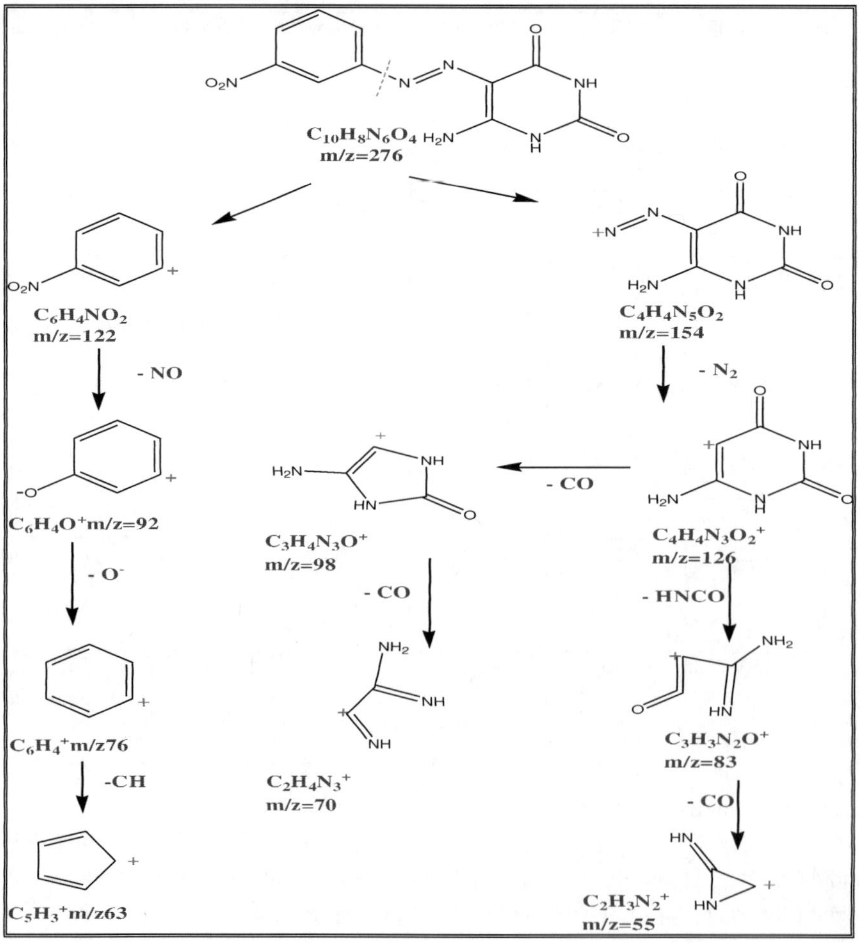

Scheme 2. Fragmentation pattern of 5-(3-nitro phenyl azo) 6-aminouracil

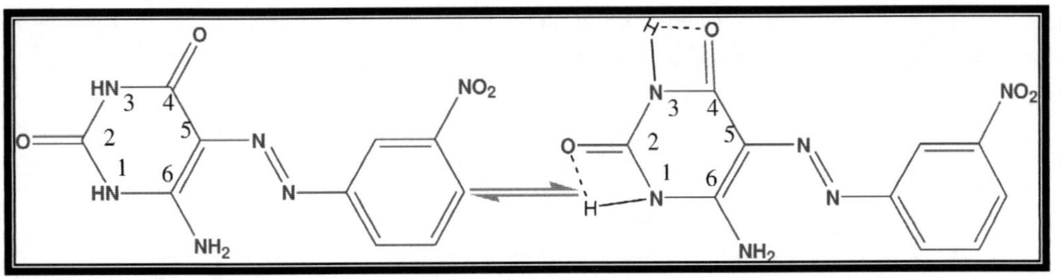

Scheme 3. Tautomer forms of 5-(m-nitro phenyl azo) 6-aminouracil ligand

The shift of bending vibration band of both NH_2 from 1532 to 1511cm^{-1} and C-NH_2 deformation from 1264 to 1216 cm^{-1} confirm that the amino NH_2 group is participated in chelation. Also, the shift in stretching vibration band of the azo group from 1449 to 1389 cm^{-1} in all complexes (except zinc complex that disappeared in it) confirm that chelation occurred through the amino NH_2 group and the azo N=N group. Two new bands within the range 576-502 cm^{-1} and 495-421 cm^{-1}observed in the spectra of the complexes and not observed in the spectrum of are ligand were due to M-O and M-N.
The IR spectra revealed that the ligand behaves as negative bidentate ligand through deprotonated of NH_2 group of uracil ring and N=N of azo group.

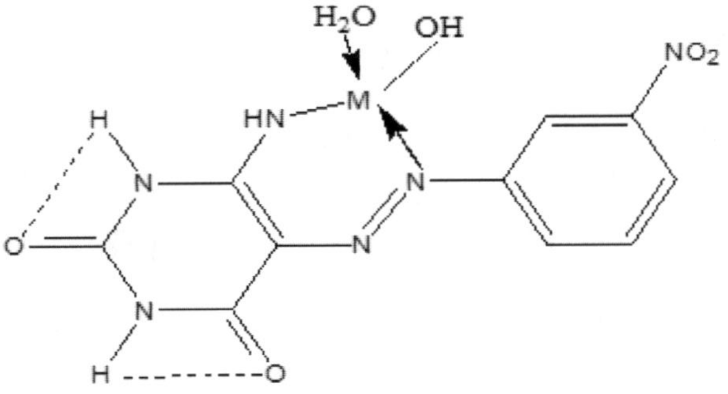

Scheme 4. Complexation pattern between metal cation and 5-(m-nitro phenyl azo) 6-aminouracil forming hydrogen bond

3.3. ^{1}HNMR spectral study

The ^{1}HNMR spectral data - listed in table 4 and represented by figure 3 show that the ligand exhibits a doublet band at 2.50 ppm assigned to the CH_3 protons of the solvent (DMSO-d^6) and a multiple band within the range 7.10 – 7.71 ppm representing the 4 CH protons of the phenyl ring of the 5-(3- nitro phenyl azo)-6-aminouracil ligand. The two NH protons in positions 1 and 3 in the uracil ring exhibited two singlet peaks at 10.61-11.01 ppm, respectively [47,48]. The two NH_2 protons in the 6 position of the uracil ring were observed at 3.17 ppm which is shifted to 3.40 ppm in the complexes due to chelation as shown in figure 5. To recognizing the exchangeable protons in the compounds, one drop of D_2O is added to the sample resulting in the disappearance of the two NH signals indicating that the protons are linked to the two oxygen atoms of the two adjacent carbonyl oxygen producing the enolic form illustrated by scheme 5 and figure 4.

The ^{1}HNMR spectra of the complexes figure 3 showed an additional singlet broad band at 3.5 ppm corresponding to the coordinated water and the OH satisfy the oxidation number on the metal ion as shown in figure 5.

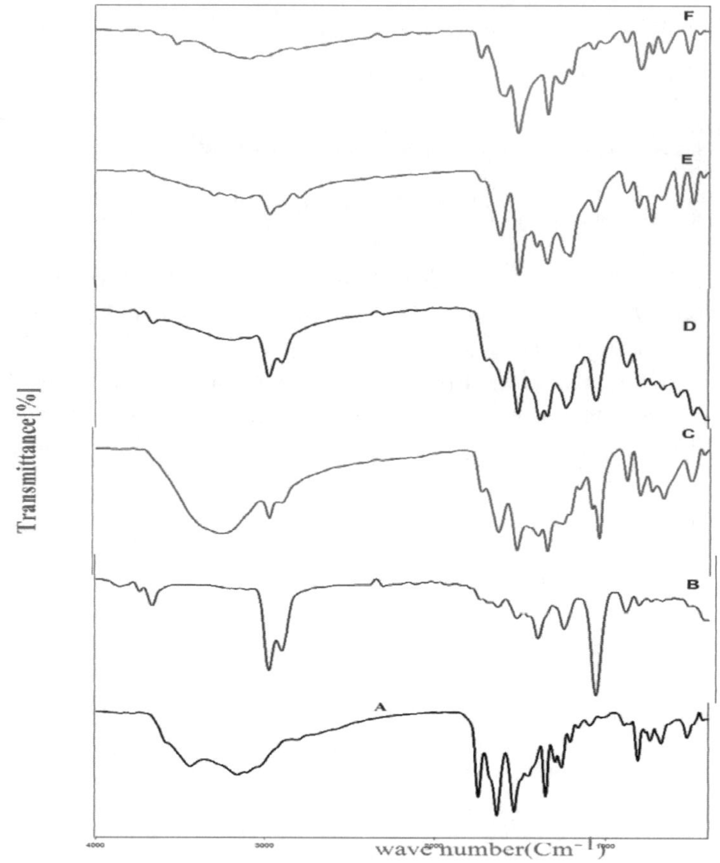

Figure 2. Infrared spectra of the 5- (3-nitro phenyl azo)6- amino uracil ligand and its prepared complexes; A---IR of ligand B----- IR of Fe-complex C-----IR of Co-complex D---- IR of Ni-complex E---- IR of Cu-complex F----- IR of Zn- complex.

The ^{1}HNMR spectra of the complexes figure 3 showed an additional singlet broad band at 3.5 ppm corresponding to the coordinated water and the OH satisfy the oxidation number on the metal ion

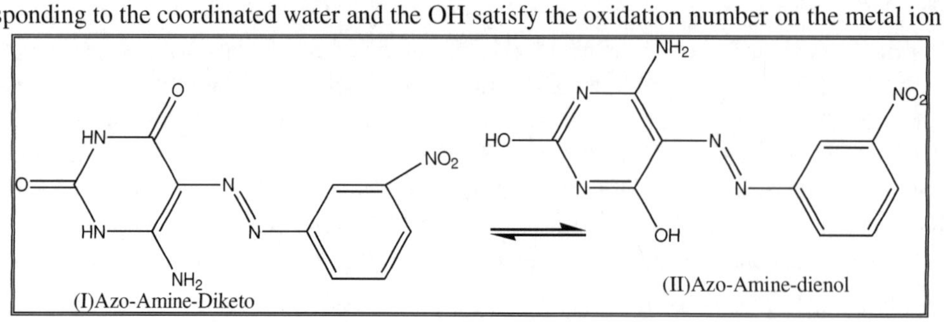

Scheme 5. Keto-enol equilibrium for 5-(3- nitro phenyl azo)-6-aminouracil ligand.

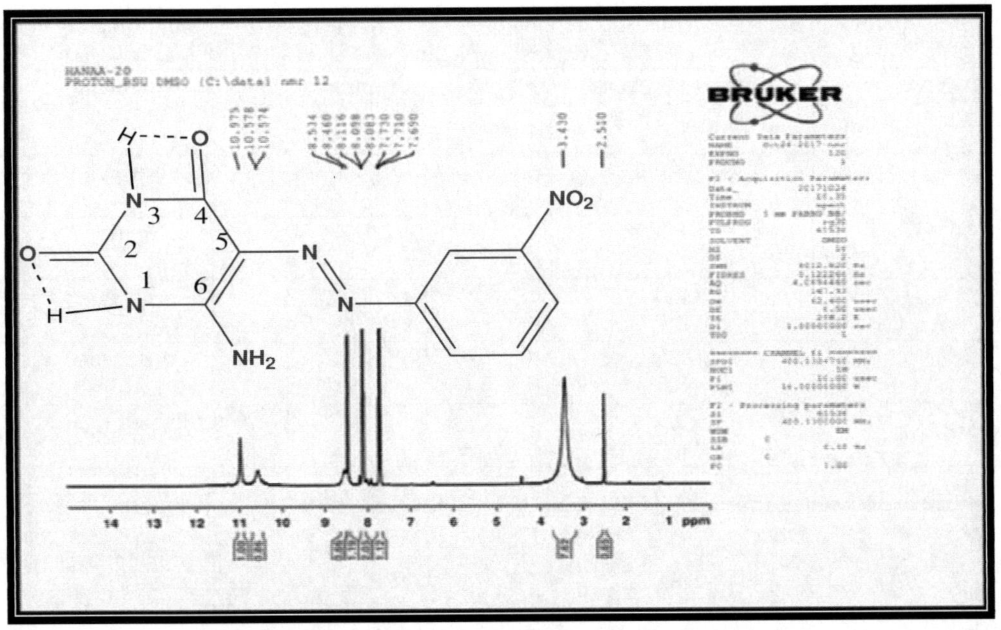

Figure 3. HNMR of 5-(3-nitro phenyl azo) 6-aminouracil

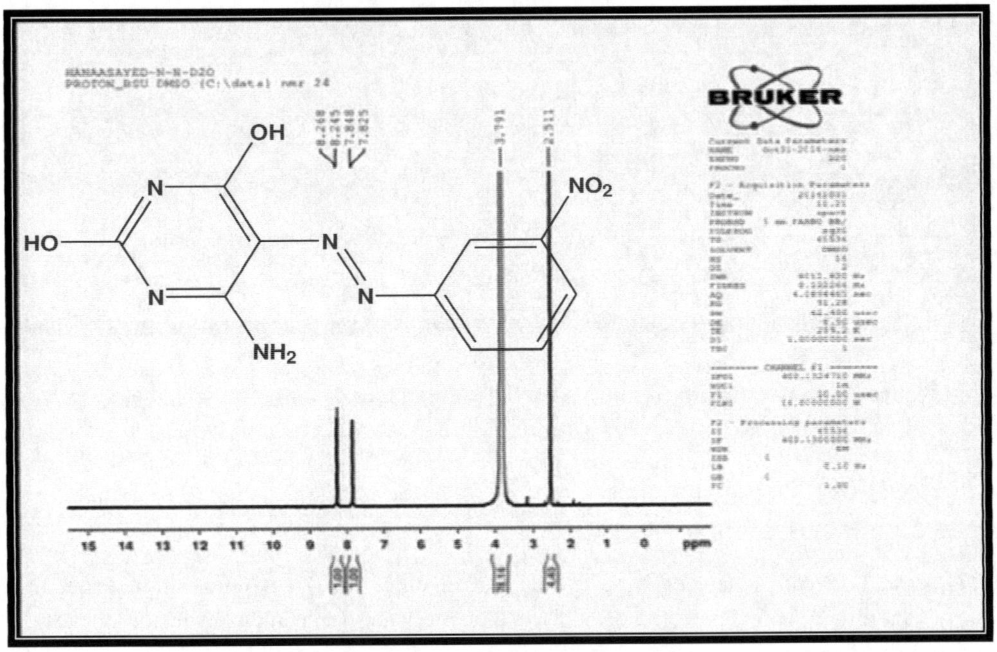

Figure 4. Duet rated NMR of 5-(m-nitro phenyl azo) 6-aminouracil

Table 4. Protocol of proton NMR spectra data of 5- (3- nitro phenylazo)-6-aminouracil ligand complex and ligand

Node	shift $_{Theoretical}$ (ppm)	Shift $_{found}$(ppm)	Comment
NH_2	8.46	3.17	Amine+ H bond
N_3H	10.00	11.01	Imide
N_1H	6.00	10.61	Urea
CH	7.26	7.10-7.71	4H Benzene
H_2O , OH	2	3.5	Coordinated H2O ,OH

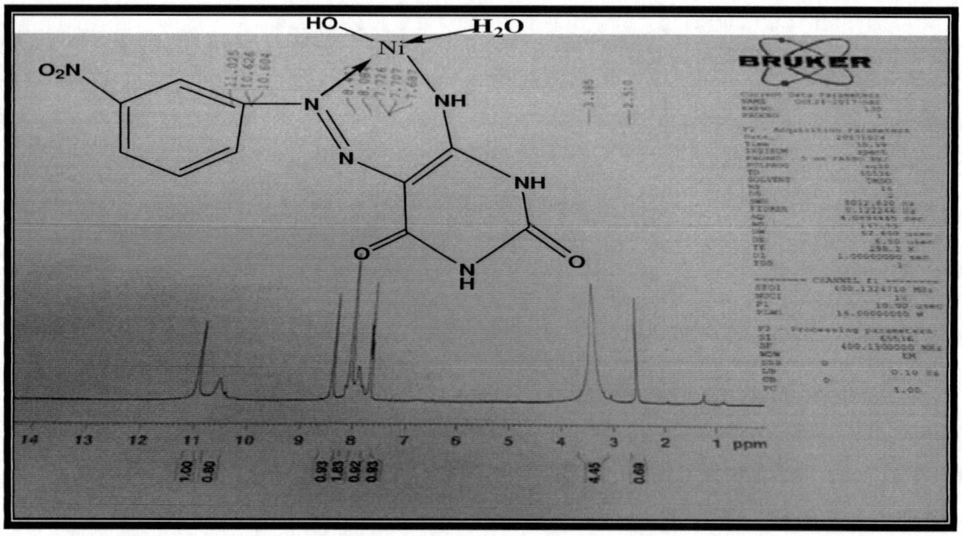

Figure 5. HNMR of Ni-5-(m-nitro phenyl azo) 6-aminouracil as represented element

3.4. Thermal analysis

The TGA and DTA of complexes were carried out in order to (i) get information about thermal stability of these new complexes, (ii) decide whether the water molecules are inside or outside the inner coordinated sphere of the central metal ion because it is impossible to differentiate between hydrated and coordinated water molecules by using IR spectral data, and (iii) suggest a general scheme for thermal decomposition of these complexes. The coordinated water molecules are more strongly bonded to the metal ion than water molecules of hydration, and hence are eliminated at higher temperature than those of water of hydration.

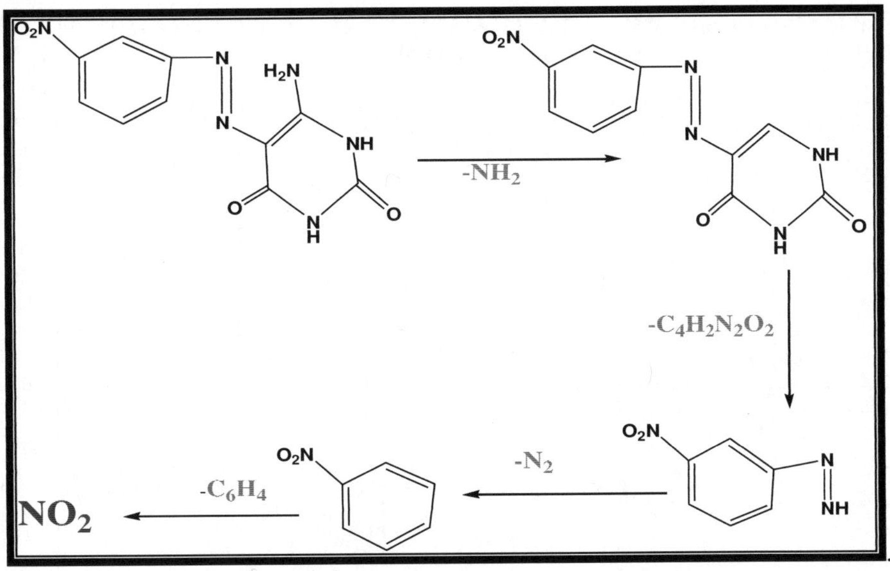

Scheme 6. Thermal degradation of 5-(3-Nitro phenylazo)-6-aminouracil

Thermal analyses data (TGA and DTA) listed in table 5 showed that 5-(3-Nitro phenylazo)-6-aminouracil undergoes an exothermic decomposition by losing the NH_2 at 147 °C. Further heating leads to the release of uracil ring at 344 °C. The azo group is released at 438 °C. The phenyl ring is released at 581 °C [27] and finally the NO_2 is released at 823 °C.

The Fe(III) complex decomposes in three phases. The first phase corresponds to the loss of a molecule of H_2O between 26–127°C with mass losses of (obs. = 4.41%, calc. = 4.49%) (52). The second phase is from 210 to 450°C and is attributed to the loss of the organic moiety $C_{20}H_{18}NS$ with mass losses of (obs. = 44.37%, calc. = 44.07%). The final phase shows the loss of the organic moiety, $C_8H_{10}OS$, at 450–700°C with mass losses of (obs. = 22.53%, calc. = 22.33%) leaving Fe_2O_3 as the final product, and the fragment C_8N is lost as $8CO_2$ and $0.5N_2$.

The decomposition of the Co(II) complex also occurred in four steps. The first step is due to the loss of a molecule of water and 2H at 41–164°C, with mass losses of (obs. = 5.72%, calc. = 5.42%). The successive decomposition occurs within a temperature range of 164–326°C and is attributed to the loss of the organic moiety CHN_2 with mass losses of (obs. = 11.12%, calc. = 10.84%). The third step is due to the loss of CH_4N_2O at 326-408°C, with mass losses of (obs. = 16.65%, calc. = 16.26%).The last step involves the loss of the organic moiety, C_6NO_2, at 408-812°C with mass losses of (obs. = 30.96%, calc. = 31.98%). The final product is CoO, and the C_5 fragment is lost as CO and CN.

The TGA curve of the Ni(II) complex reveals a two-step decomposition. The first is the loss of two molecules of water at 50–191°C, with mass losses of (obs. = 9.11%, calc. = 8.89%). The second step ranges from 288 to 484°C and is assigned to the loss of the organic moiety, $C_{10}H_9N_6O_5$with mass losses (obs. = 72.44%, calc. = 72.64%). The overall weight loss (81.55%) calc and (81.57%) found leaving NiO as residue.

Table 5 (a). Thermal analyses data (TG-DTA) of 5-(3-nitro phenylazo)-6-aminouracil and its complexes

Compound	Temp. range [^oC]	Tm [^oC]	Peak type	Calc. weight loss [%]	found weight loss [%]	Losing fragment
Ligand	103- 196	345	Exothermic	5.80	6.01	NH_2
	280 -400		Exothermic	39.85	39.80	$C_4H_2N_2O_2$
	403-493		Exothermic	10.15	10.28	N=N
	497-697		Exothermic	27.17	27.23	C_6H_5
	800-936		Exothermic	16.67	16.14	NO_2
$[Fe(H_2L)(OH)_2(H_2O)_2]$	26-127		Endothermic	4.49	4.41	H_2O
	127-316	308	Exothermic	18.21	18.43	$C_2H_5N_2O$
	316-451		Exothermic	19.20	19.11	C_5H_3N
	451-564		Endothermic	9.48	9.63	C_2N
	564-1096		Endothermic	19.70	20.06	$H_3N_2O_3$
			Endothermic	28.89	28.31	Residue 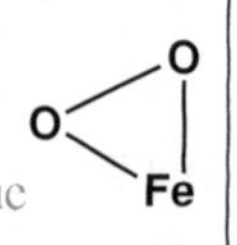 + CO

Table 5 (b). Thermal analyses data (TG-DTA) of 5-(3-nitro phenylazo)-6-aminouracil and its complexes

$[CoH_2L(OH)(H_2O)]$	41-164	351	Endothermic	5.42	5.72	H_2O +2H
	164-326		Exothermic	10.84	11.12	CHN_2
	326-408			16.26	16.55	CH_4N_2O
	408-812		Exothermic	31.98	30.96	C_6HNO_2
			Exothermic	34.94	34.57	Residue **[Co O+** CO+CN]
$[NiH_2L(OH)(H_2O]2H_2O$	46-191	416	Weak endothermic	8.89	9.11	Dehydration of 2 H_2O
	288-484		Strong endothermic	72.64	72.42	$C_6H_4NO_2$ $+C_4H_3N_3O_2+H_2O$
				18.45	18.47	Residue **Ni O**

Cu(II) complex decomposes in four steps. The first step is attributed to the loss of the two molecules of water, with mass losses of (obs. = 9.35%, calc. = 8.79%) at 13–201°C. The second step ranges from 201 to 300°C and is attributed to the loss of the fragment H_2O and 2H, with mass losses of (obs. = 5.00%, calc. = 4.88%). The third step ranges from 300 to 617°C and is attributed to the loss of the organic moiety $C_4H_2N_3O_2$, with mass losses of (obs. = 31.88%, calc. = 30.27%).The final step is from 617 to 1101°C corresponding to the loss of the CHN_2, with mass losses of (obs. = 10.78%, calc. = 10.49%). The remaining residue is mixture of(CuO and $C_4H_3NO_2$.

Table 5 (c). Thermal analyses data (TG-DTA) of 5-(3-nitro phenylazo)-6-aminouracil and its complexes

$[Cu(H_2L)(OH)(H_2O)]2H_2O$	13-201	397	Endothermic	8.79	9.35	$2H_2O$
	201-300		Exothermic	4.88	5.00	H_2O+2H
	300-617			30.27	31.88	$C_4H_2N_3O_2$
	617-1101		Exothermic	10.49	10.72	CHN_2
			Exothermic			Residue
				43.10	43.32	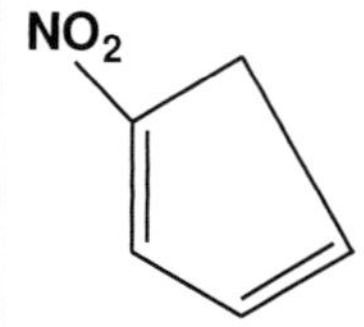 **+ Cu O**
$[ZnH_2L(OH)(H_2O)]$	34-168	394	Endothermic	4.79	3.71	H_2O
	168-340		Exothermic	7.99	7.92	N_2+2H
	340-550		Endothermic	29.30	29.44	$C_4H_2N_2O_2$
	550-1101		Endothermic	36.22	36.20	$C_6H_4N_2O_2$
						Residue
				22.21	22.71	**ZnO**

The Zn(II) complex decomposes in four steps. Step one is between 34–168°C, which indicates the loss of H_2O, with mass losses of (obs. = 3.71%, calc. = 4.79%). The second step involves the loss of azo group and 2H from 168 to 340°C, with mass losses of (obs. = 7.92%, calc. = 7.99%). The third step involves the loss of the organic moiety $C_4H_2N_2O_2$, from 340 to 550°C, with mass losses of (obs. = 29.44%, calc. = 29.30%). The final step is attributed to the loss of the organic moiety, $C_6H_4N_2O_2$, at 550–1101°C, with mass losses of (obs. = 36.20%, calc. = 36.22%), leaving behind the ZnO residue and the fragment C_3 is lost as $3CO_2$.

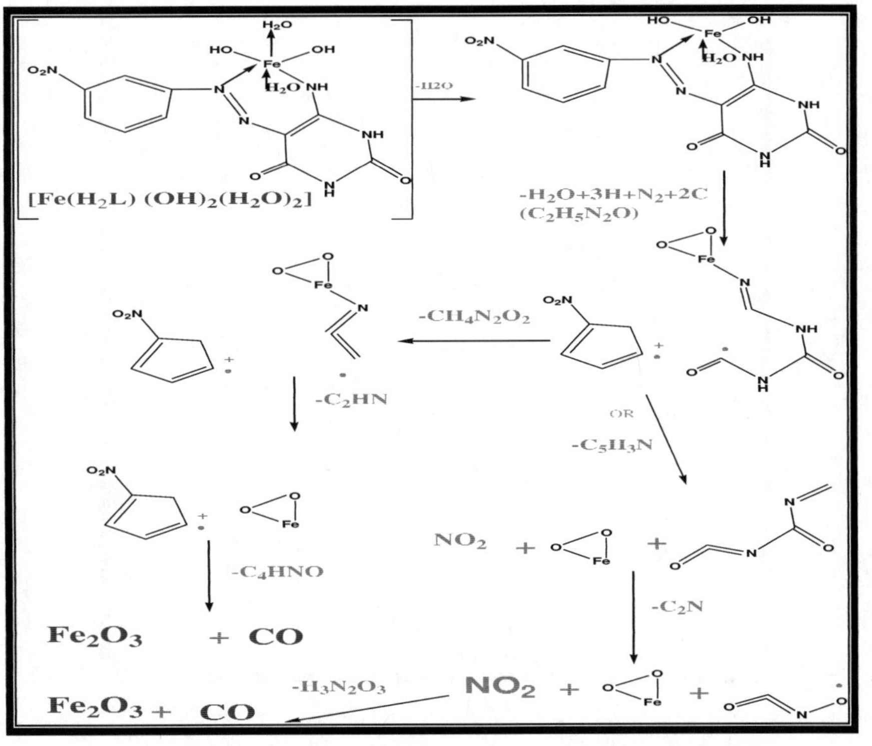

Scheme 7. Thermal degradation of Fe (III) complex.

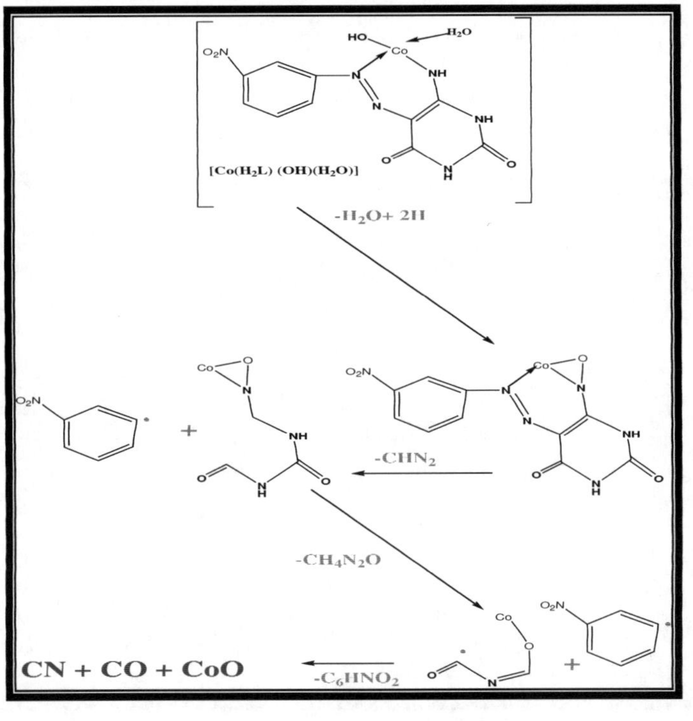

Scheme 8. Thermal degradation of Co(II) complex.

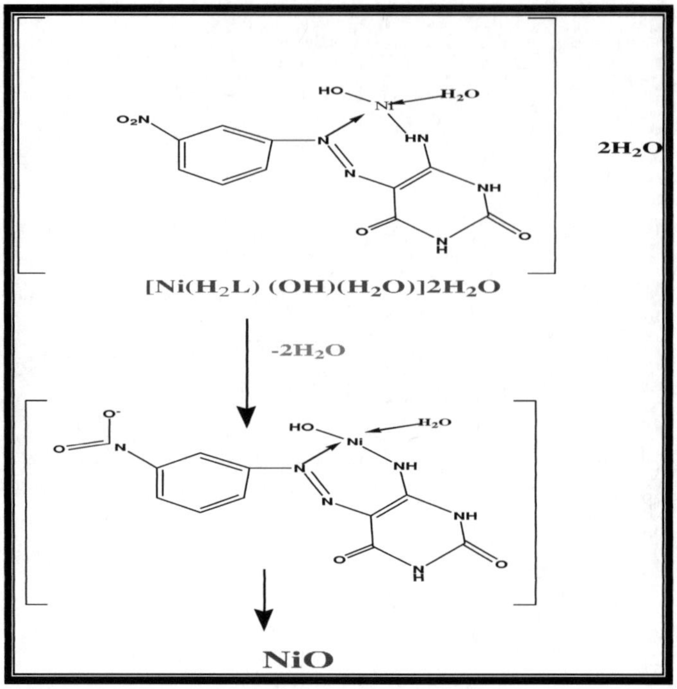

Scheme 9. Thermal degradation of Ni (II) complex.

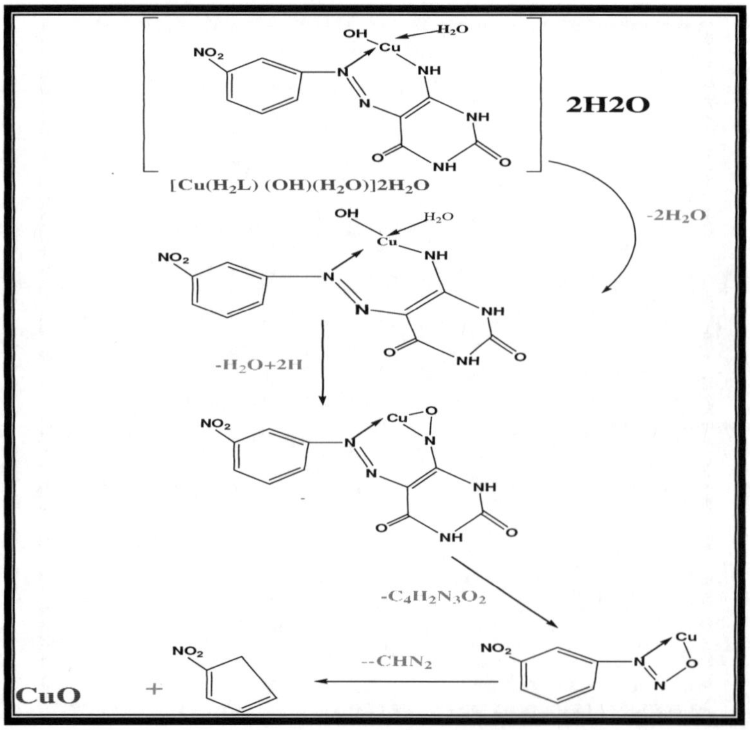

Scheme 10. Thermal degradation of Cu (II) complex.

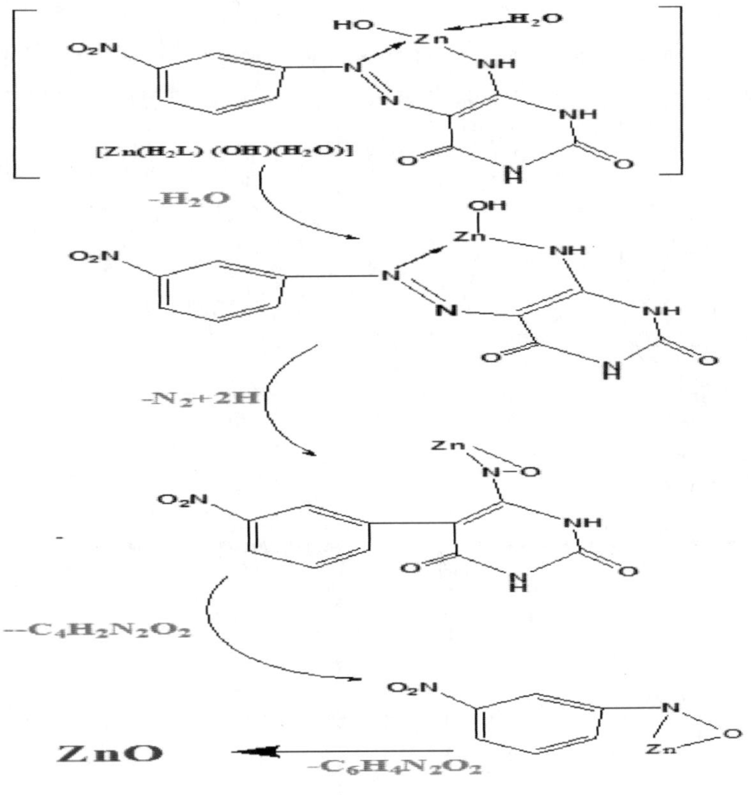

Scheme 11. Thermal degradation of Zn (II) complex.

In all cases with the exceptions of the ligand and the Cu(II) complex, the decomposition pattern showed the loss of carbon fragments which got oxidized to CO_2, and hydrogen or nitrogen which were lost as gases. Thus, the decomposition pattern corroborates the proposed formulation of the complex. formulation of the complex.

3.5. Magnetic moment measurements

On the basis of the magnetic moment measurements lifted in Table 6 it is found that all the complexes under study are paramagnetic except for the diamagnetic zinc complex that shows square planar geometry because of the exact value of magnetic moment depends on the temperature of measurement and the magnitude of the spin-orbit coupling, all measurements were carried out at room temperature. In the present investigation, Fe^{+3} complex has magnetic moment of about 2.3 [B.M] indicating low-spin octahedral state which corresponds to the presence of one unpaired electron.
Co^{+2} complexe has a magnetic moment value of 4.86BM at room temperature, suggesting tetrahedral geometry for this complex which is confirmed by the presence of 3 unpaired electrons. The observed magnetic moment value for $Ni^{+2}$is 3.75 (BM) which falls in the range of tetrahedral and square planar structure for this complex but the presence of 2 unpaired electrons excludes Square planar geometry and confirms tetrahedral structure for the complex.
The Cu^{+2} complex, under investigation showed magnetic moment of 2.4[BM] which suggested tetrahedral structure that supported by thermal analysis of this complex. The relatively high magnetic moment value can be attributed to ferromagnetism on the complex [49–54].

Table 6. The magnetic susceptibility measurements, diamagnetic corrections and the effective magnetic moments of Fe^{+3}, Co^{+2}, Ni^{+2}, $Cu^{+2}$and Zn^{+2} complexes:

Complex	Xg *10^{-5} Cgs unite	Xm*10^{-3} Cgs unite	XD *10^{6} Cgs unite	X'm *10^{-3} Cgs unite	M_{eff}	M_{eff} Theo	Nomber of unpaired e	Geometry
$[Fe(H_2L)(OH)_2(H_2O)_2]$	5.5468	2.2234	12.8	2.2106	2.30	2-2.5	1	Octahedral(low spin)
$[CoH_2L(OH)(H_2O)]$	2.5032	9.2350	12.8*	9.2222	4.688	3.87	3	Tetrahedral
$[NiH_2L(OH)(H_2O]2H_2O$	1.4602	5.9092	12.8*	5.8964	3.75	2.83	2	Tetrahedral
$[Cu(H_2L)(OH)(H_2O)]2H_2O$	5.95	2.4375	12.8*	2.4247	2.40	1.7-2.2	1	Tetrahedral
$[Zn(H_2L)(OH) (H_2O)]$	4.542	1.705	13	1.692	0	1.7	0	Square planar

X_M = Molar susceptibility
X_D = Diamagnetic susceptibility
X'M = Corrected Molar susceptibility
M_{eff} = Effective Magnetic moment

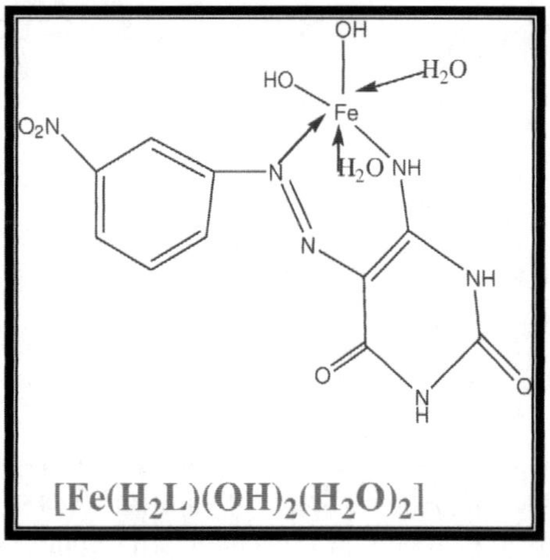

Scheme 12. Geometry of Fe-5-(m-nitro phenyl azo) 6-aminouracil complex

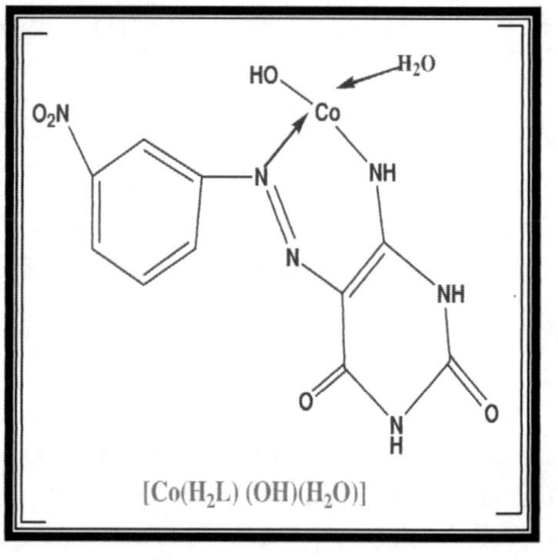

Scheme 13. Geometry of Co-5-(m-nitro phenyl azo) 6-aminouracil complex

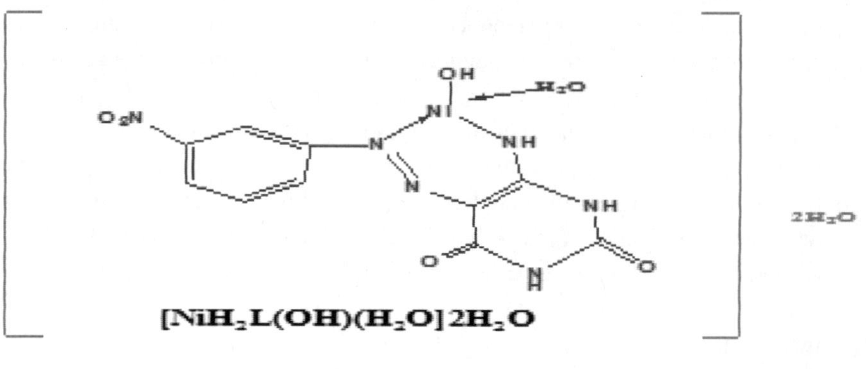

Scheme 14. Geometry of Ni-5-(m-nitro phenyl azo) 6-aminouracil complex

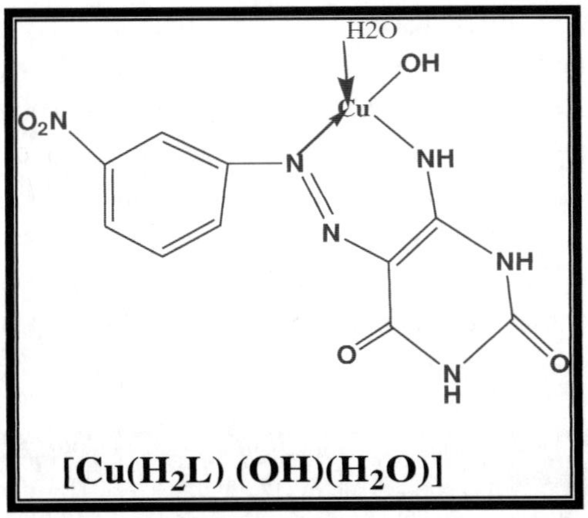

Scheme 15. Geometry of Cu-5-(m-nitro phenyl azo) 6-aminouracil complex

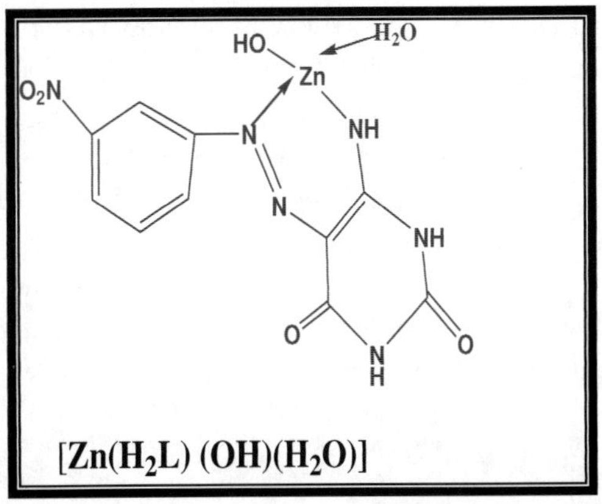

Scheme 16. Geometry of Zn-5-(m-nitro phenyl azo) 6-aminouracil complex

3.6. The antibacterial assay

Table 7 shows the diameters of the inhibition zones of the different bacterial strains against gram positive and gram negative bacteria. Also Figure 6 shows the inhibition zone (mm) of the cobalt and Zinc complexes. Besides the anti-microbial activity increased against gram positive indicated with increased zone of inhibition than negative one but the showed also activity against gram negative with anti-fungal activity.

Table 7. The diameters of the zones against different bacterial strains

sample / Microorganism	**Ligand**	**Ni Complex**	**Fe Complex**	**Cu Complex**	**Zn Complex**	**Co Complex**	Standard antibiotic
Gram negative bacteria	Gentamicin						
Escherichia coli	NA	22.6±0.5	NA	11±0.5	27	25	34±0.1
Klebsiella pnemonia	*NA*	NA	NA	NA	NA	27	45±0.5
Pseudomonas aeruginosa	NA	NA	NA	NA	-	-	35±0.4
Gram positive bacteria	Ampicillin						
Staphylococcus aureus	NA	NA	NA	NA	NA	NA	33±0.2
Streptococcus mutans.	NA	9.6±0.5	NA	NA	29±1	27.6±0.5	22±0.1
Fungi	Nystatin						
candida albicans	NA	15.3±0.5	NA	NA	NA	17.3±0.5	28±0.2

The current research studied the antibacterial activity of different complexes on different bacterial strains besides their antifungal activities. Especially both cobalt and zinc nanocomposities showed good antibacterial activity against gram-positive and gram-negative bacteria. Regarding the cobalt antibacterial activity that's might be attributed to the cobalt nanomaterial have the ability for cell membrane attachment and disruption so increase the bacterial permeability with bactericidal activity. In addition cobalt in nano size had the ability for interacting with the sulfur- and phosphorus-containing compounds in the bacterial cell lead to lose the bacterial activity. Besides cobalt in nano form can interact with the DNA of the bacterial cells causes' damage in the DNA and stop its replications besides bacterial cell death. Cobalt ion itself when released from the nano-composites causes severe antibacterial activity [55].

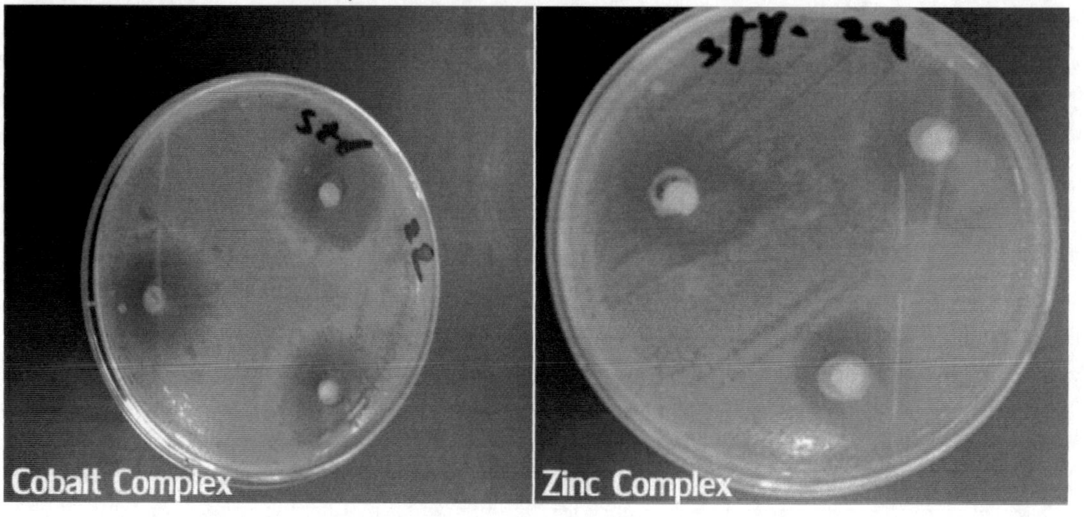

Figure 6. Shows the Zone of inhibition (mm) of both cobalt and zinc complexes against streptococcal infection

Zinc complex also showed good antibacterial activity in this study which might be attributed to the presence of the water molecules which responsible for generating free radicals like hydroxyl or oxygen species that responsible for the oxidative stress and antimicrobial activity.

4. Conclusion

5-(3-nitrophenyllazo)-6-minouracil behaves as monobasic bi dentate ligand that bonds to the metal ion through both NH_2 group and N=N groups forming six membered ring. These complexes have high thermal stability and are biologically active towards bacteria gram (-ve), gram (+ ve) and fungi.

References

[1] Seferoglu Z and Ertan N 2007 Synthesis and spectral properties of new hetarylazo indole dyes *Russ. J. Org. Chem.* **43** 1035–41

[2] Seferoğlu Z and Ertan N 2007 Synthesis of some novel bis(hetaryl)azo disperse dyes and investigation of their absorption spectra *Heteroat. Chem.* **18** 622–30

[3] Y.H.MARIUM R B M and 1973*Metal Ions in Biological Systems: Volume 8*

[4] Sujamol M S, Athira C J, Sindhu Y and Mohanan K 2010 Synthesis, spectroscopic characterization, electrochemical behaviour and thermal decomposition studies of some transition metal complexes with an azo derivative *Spectrochim Acta A Mol Biomol Spectrosc* **75** 106–12

[5] Diab M A, El-Bindary A A, El-Sonbati A Z and Salem O L 2012 Supramolecular structure and substituents effect on the spectral studies of dioxouranium(VI) azodyes complexes *J. Mol. Struct.* **1007** 11–9

[6] Hassanzadeh A, Zeini-Isfahani A, Habibi M H, Heravi M R A P and Abdollahi-Alibeik M 2006 1H, 13C, NH, HH, CH COSY, HH NOESY NMR and UV–vis studies of Solophenyl red 3BL dye azo-hydrazone tautomerism in various solvents *Spectrochim. Acta Part A Mol. Biomol. Spectrosc.* **63** 247–54

[7] Pirillo S, Einschlag F S G, Ferreira M L and Rueda E H 2010 Eriochrome Blue Black R and Fluorescein degradation by hydrogen peroxide oxidation with horseradish peroxidase and hematin as biocatalysts *J. Mol. Catal. B Enzym.* **66** 63–71

[8] Pirillo S, Rueda E H and Ferreira M L 2012 Supported biocatalysts for Alizarin and Eriochrome Blue Black R degradation using hydrogen peroxide *Chem. Eng. J.* **204–205** 65–71

[9] Rageh N M 2004 Tautomeric structures, electronic spectra, acid–base properties of some 7-aryl-2,5-diamino-3(4-hydroxyphenyazo)pyrazolo[1,5-a]pyrimidine-6-carbonitriles, and effect of their copper(II) complex solutions on some bacteria and fungi *Spectrochim. Acta Part A Mol. Biomol. Spectrosc.* **60** 1917–24

[10] Li X, Wu Y, Gu D and Gan F 2010 Spectral, thermal and optical properties of metal(II)–azo complexes for optical recording media *Dye. Pigment.* **86** 182–9

[11] Masoud M S, Ibrahim A A, Khalil E A and El-Marghany A 2007 Spectral properties of some metal complexes derived from uracil–thiouracil and citrazinic acid compounds *Spectrochim. Acta Part A Mol. Biomol. Spectrosc.* **67** 662–8

[12] Bayrak C 2012 Vibrational spectroscopic study of 6-aminouracil tetracyanonickelate complexes *Hacettepe J. Biol. Chem* **40** 419–26

[13] Edupuganti R, Wang Q, Tavares C D J, Chitjian C A, Bachman J L, Ren P, Anslyn E V and Dalby K N 2014 Synthesis and biological evaluation of pyrido[2,3-d]pyrimidine-2,4-dione derivatives as eEF-2K inhibitors *Bioorg. Med. Chem.* **22** 4910–6

[14] Pałasz A and Cież D 2015 In search of uracil derivatives as bioactive agents. Uracils and fused uracils: Synthesis, biological activity and applications *Eur. J. Med. Chem.* **97** 582–611

[15] El-Barbary A A, Hafiz Y A and Abdel-Wahed M S 2011 Synthesis, characterization, and biological activity of some aza-uracil derivatives *J. Heterocycl. Chem.* **48** 639–44

[16] Holm R H, Kennepohl P and Solomon E I 1996 Structural and Functional Aspects of Metal Sites in Biology *Chem. Rev.* **96** 2239–314

[17] Akbar Ali M and Livingstone S E 1974 Metal complexes of sulphur-nitrogen chelating agents *Coord. Chem. Rev.* **13** 101–32

[18] Singleton P 1999 *Bacteria in Biology, Biotechnology and Medicine* vol 5th ed.

[19] Tenaillon, Olivier; Skurnik, David; Picard, Bertrand; Denamur E 2015 The population genetics of commensal Escherichia coli *Nat. Rev. Microbiol.* **8** 207–17

[20] Masalha M, Borovok I, Schreiber R, Aharonowitz Y and Cohen G 2001 Analysis of transcription of the Staphylococcus aureus aerobic class Ib and anaerobic class III ribonucleotide reductase genes in response to oxygen *J. Bacteriol.* **183** 7260–72

[21] Ghosh P, Mukhopadhyay T K and Sarkar A R 1984 Interaction of divalent metal ions with uracil III. Complexes of MnII, FeII, CoII, NiII and CuII with uracil acting as bidentate ligand *Transit. Met. Chem.* **9** 46–8

[22] Ali O Y and Fridgen T D 2011 Structures of electrosprayed Pb(Uracil-H)+ complexes by infrared multiple photon dissociation spectroscopy *Int. J. Mass Spectrom.* **308** 167–74

[23] Narang K K, Singh V P and Bhattacharya D 1997 Synthesis, characterization and antitumour activity of uracil and uracil–histidine complexes with metal(III) ions *Transit. Met. Chem.* **22** 333–7

[24] Singh V P, Singh A, Narang K K and Bhattacharya D 2004 5-Bromouracil and 5-bromouracil–histidine complexes with metal(III) ions and their antitumour activity *Transit. Met. Chem.* **29** 107–11

[25] Singh V P, Singh S, Narang K K and Bhattacharya D 2009 Aluminium(III), chromium(III) and iron(III) complexes with 5-iodouracil and 5-iodouracil-histidine and their antitumour

activity *J. Enzyme Inhib. Med. Chem.* **24** 105–10

[26] Moanta A, Ionescu C, Tutunaru B and Dumitru M 2010 Thermal and electron impact decomposition of 4-hydroxy-4′-cyano- azobenzene *Rev. Chim.* **61**

[27] Schwarzenbach G 1957 *Complexometric titrations.* vol 89 (Methuen Co., London)

[28] Vogel A I 1961 *A text book of Quantitative Inorganic Analysis, including elementary instrumental analysis*

[29] Collee JG, Miles RS W B 2006 *Tests for identification of bacteria. In Collee JG, Fraser AG, Marmion BP, Simmons A. eds. Mackie and McCartney's Practical medical microbiology, 14th ed* (Churchill Livingstone)

[30] Wiegand I, Hilpert K and Hancock R E W 2008 Agar and broth dilution methods to determine the minimal inhibitory concentration (MIC) of antimicrobial substances *Nat. Protoc.* **3** 163–75

[31] C.Scott A 1989 *Laboratory control of antimicrobial therapy. In: Collee JG et al. eds. Practical Medical Microbiology, 13th Edition* (Edinburgh: Churchill Livingstone)

[32] Bernier J-L and Hénichart J-P 2018 Electron impact fragmentation of 6-aminouracils *Biomed. Mass Spectrom.* **10** 626–8

[33] Moanta A 2012 Characteristic Fragmentation Patterns of Some 4- (phenylazo) phenols Obtained by Electron Impact Mass Spectrometry *Rev. Chim.* **63** 7–9

[34] Sadr-Arani L, Mignon P, Chermette H, Abdoul-Carime H, Farizon B and Farizon M 2015 Fragmentation mechanisms of cytosine, adenine and guanine ionized bases *Phys. Chem. Chem. Phys.* **17** 11813–26

[35] Issa Y M, Hassib H B and Abdelaal H E 2009 1H NMR, 13C NMR and mass spectral studies of some Schiff bases derived from 3-amino-1,2,4-triazole *Spectrochim. Acta - Part A Mol. Biomol. Spectrosc.* **74**

[36] Curtis N F 1972 Some metal-ion complexes with ligands formed by reaction of amines with aliphatic carbonyl compounds. Part I. Nickel(II) and copper(II) complexes formed by the diaminoethane–acetone reaction *J. Chem. Soc. Dalt. Trans.* 1357–61

[37] Ruiz-Sánchez J, Colacio-Rodríguez E, Salas-Peregrin J M and Romero-Molina M A 1986 Thermal decomposition of 6-amino-1,3-dimethyl-5-phenylazouracil complexes of Co(II), Ni(II), Cu(II) and Ag(I) *J. Anal. Appl. Pyrolysis* **9** 159–70

[38] Masoud M S, Abou El-Enein S A, Ayad M E and Goher A S 2004 Spectral and magnetic properties of phenylazo-6-aminouracil complexes *Spectrochim. Acta Part A Mol. Biomol. Spectrosc.* **60** 77–87

[39] Yadav R A, Yadav P N S and Yadav J S 1988 Vibrational studies of biomolecules—II. 2-thiocytosine *Spectrochim. Acta Part A Mol. Spectrosc.* **44** 1201–6

[40] Durig J R 1993 *Vibrational Spectra and Structure* vol 20 (Elseiver science publishers)

[41] Colombo L and Kirin D 1986 Raman spectrum of 1-methyl-uracil single crystal. Interpretation of internal and external spectra *Spectrochim. Acta Part A Mol. Spectrosc.* **42** 557–65

[42] Debnath D, Roy S, Li B-H, Lin C-H and Misra T K 2015 Synthesis, structure and study of azo-hydrazone tautomeric equilibrium of 1,3-dimethyl-5-(arylazo)-6-amino-uracil derivatives *Spectrochim. Acta Part A Mol. Biomol. Spectrosc.* **140** 185–97

[43] Krause R A and Krause K 1980 Chemistry of bipyridyl-like ligands. Isomeric complexes of ruthenium(II) with 2-(phenylazo)pyridine *Inorg. Chem.* **19** 2600–3

[44] Masoud M S and Refaat L S 1982 Synthesis and characterization of some nickel(II) Schiff bases complexes *Transit. Met. Chem.* **7** 315–8

[45] Ahmed F, Dewani R, Pervez M K, Mahboob S J and Soomro S A 2016 Non-destructive FT-IR analysis of mono azo dyes *Bulg. Chem. Commun.* **48** 71–7

[46] Larkin P J 2011 *Infrared and Raman spectroscopy : principles and spectral interpretation*

[47] Seferoğlu Z and Ertan N 2008 Synthesis, characterization and spectroscopic properties of some new phenylazo-6-aminouracil *Cent. Eur. J. Chem.* **6** 81–8

[48] Seferoğl Z 2009 A study on tautomeric equilibria of new hetarylazo-6-aminouracils *Arkivoc* **2009** 42–57

[49] Lewis J and Wilkins R G 1960 Modern Coordination Chemistry: Principles and Methods

[50] Angelici R J 1977 *Synthesis and Technique in Inorganic Chemistry (Saunders golden sunburst series)* (W.B. Saunders Company)
[51] Sharpe C E H and A G 2008 *Inorganic Chemistry* (Pearson Education Ltd, Harlow)
[52] Evans D F 1959 400. The determination of the paramagnetic susceptibility of substances in solution by nuclear magnetic resonance *J. Chem. Soc.* 2003
[53] Szafran Z, M.Pike R and M. Singh M 1991 *Microscale Inorganic Chemistry- A comprehensive laboratory experience*
[54] Mcmills L, Nyasulu F and Barlag R 2014 Magnetic Susceptibility of Coordination Compounds in the General Chemistry Laboratory *J. Lab. Chem. Educ.* **2** 11–4
[55] El-Shahawy A A G, Abo El-Ela F I, Mohamed N A, Eldine Z E and El Rouby W M A 2018 Synthesis and evaluation of layered double hydroxide/doxycycline and cobalt ferrite/chitosan nanohybrid efficacy on gram positive and gram negative bacteria *Mater. Sci. Eng. C* **91** 361–71

Facile sonochemical synthesis of artichoke- like Co_3O_4-graphene nanocomposites

E M Kamar[1] and S I El-Dek[2]*

[1]Chemistry Department, Faculty of Science, Benha University, Benha, Egypt
[2]Materials Science and Nanotechnology Dept., Faculty of Post graduate studies for Advanced Sciences, Beni-Suef University, Beni-Suef, Egypt.
Corresponding author e-mail: samaa@psas.bsu.edu.eg , didi5550000@gmail.com.

Abstract. Facile sonochemical technique was used for the first time to synthesize Co_3O_4 -graphene nanocomposites at room temperature. The nanocomposites under investigation were prepared with different Co_3O_4 molar ratios of nanoparticles. The samples were characterized by X-ray diffraction (XRD), field emission scanning electron microscope (FESEM), Fourier transform infrared spectroscopy (FT-IR) and UV-vis spectroscopy. The results show that Co_3O_4 nanoparticles were distributed uniformly on the surface of graphene sheet. This work reports a simple preparation approach and recommends the described methodology for the decoration of graphene sheet surface with miscellaneous metal oxide nanoparticles.

Keywords: Graphene; Co_3O_4 nanoparticles; nanocomposite; morphology; FESEM

1. Introduction

Graphene is one of the most attractive materials being widely researched in recent times. The increased popularity of graphene originates from its wondrous properties, which include, but are not limited to, excellent electron mobility through its atom-thick sp^2 bonded 2-D structure, its large current density, optical transmittance, very good mechanical properties and remarkable thermal conductivity [1–5]. Significant synergistic effects often occur in graphene/metal oxide nano composites owing to size effects and interfacial interactions.

The synergistic effects in graphene/metal oxide composites can be underlined as follows: (i) graphene is a novel 2D support for homogeneously nucleating, growing or assembling metal oxides with distinguished size, shape, and crystallinity; (ii) metal oxides amongst the layers of graphene can proficiently defeat the re-stacking of graphene; (iii) graphene can act as a 2D conductive model for construction of a 3D interlocked conductive porous network to enhance both electrical conductivity and charge transportation of pure oxides; (iv) graphene can suppress the volume change and particle agglomeration of metal oxides during the charge–discharge process; (v) oxygen-containing functional groups on graphene guarantees respectable interfacial interactions as well as electrical contacts between graphene and metal oxides. Due to these good special effects, integration of metal oxides nanoparticles and graphene in a composite fully uses each active component and subsequently realizes outstanding electrochemical performance.
The potential applications of metal oxide–graphene nanocomposites include super capacitors, anodes for Li-ion batteries, electrocatalysts, and biosensors where several structural models of graphene-cobalt oxide composites are already proposed [6–12].

Herein, we report a facile strategy to synthesize nano composite of Co_3O_4 anchored on conducting graphene as an advanced anode for Li-ion batteries. The objective of this work was to prepare and to characterize the (G-Co_3O_4) nanocomposites in different proportions in order to verify their physicochemical properties. Moreover, one of our goals is to explore the effect of Co_3O_4 weight ratio on the properties of nanocomposites to highlight the applications of such new nanocomposites and to open a new era for another potential uses.

2. Materials and method

2.1 Synthesis of Co_3O_4 nanoparticles

All chemical reagents (analytical grade) were used as received without further purification. Deionized water was used throughout this study.

Simply, an aqueous solution of ammonium carbonate (1.86 g) was added to a stirring solution of Co $(CH_3COO)_2.4H_2O$; (0.8g). The reaction was transferred into a Teflon lined stainless steel autoclave which was hydrothermally treated at 160 °C for 24 hrs.

The pink precipitate of $CoCO_3$ was thoroughly washed and dried at 60°C for 24h. The obtained powder was calcined in air at 300 °C for 2 h in air to obtain Co_3O_4 nanoparticles [13].

2.2 Synthesis of Co_3O_4- graphene nanocomposites

Nanocomposites of Co_3O_4-graphene were prepared using modified sonochemical technique. The details were as follows: Preparation of graphene sheets (G) using graphene powder dispersed in deionized water using ultrasonic bath at a power 700W for 1h to fully exfoliate the grapheme until a homogeneous sheet was observed. All nanocomposites were prepared by adding 5, 10 and 20% of Co_3O_4 directly into 20mL of the dispersed graphene using ultrasonication for 90 min, then dried at 40 °C in air for 48hrs.

2.3 Characterizations

X-ray diffraction (XRD) patterns were examined using Bruker X-ray powder diffractometer (model D8 Advance) with (Cu $K\alpha_1$, λ=1.5417 Å). The morphology of the nanocomposites was characterized by field emission scanning electron microscopy (FESEM, Model Quanta 250 FEG). Fourier transform infrared spectroscopy (FT-IR) were obtained using Bomem; model MB1575. The UV-vis spectra were recorded on a UV-vis spectrophotometer (Jasco; model 530).

3. Results and discussion

3.1 XRD diffraction

The XRD patterns of graphene, Co_3O_4, and the as-prepared nanocomposites (Co_3O_4-graphene) were illustrated in figure 1 (a-e) For graphene, a very sharp peak was observed at 26.52°, which is attributed to the C (002) plane [14,15]. The data were indexed according to the reported pattern. of the ICDD card 08-0415 which assures the hexagonal crystal symmetry belonging to the $P6_3/mmc$ space group.

Co_3O_4 nanoparticles were crystallized in single phase cubic spinel structure as compared and indexed with ICDD card no (78-1970). All peaks appear to be broad with small intensities pointing to nanoscale particles. All planes concerning the Co_3O_4 were indexed on the basis of the cubic spinel structure of space group Fd3m. The key planes in the XRD pattern of pure Co_3O_4 were assigned to (111), (220), (311), (400), (422), (511) and (440) respectively.

The lattice parameter was computed based on cubic structure and is a= 8.0804 Å. The theoretical density was calculated from D_x = ZM/NV, where Z counts for the number of molecules per unit cell (Z=8), M is the molecular weight, N is Avogadro's number and V is the unit cell volume. It was found to be =6.063g/cm^3 which agree well with that reported in the corresponding ICDD card. The X-ray

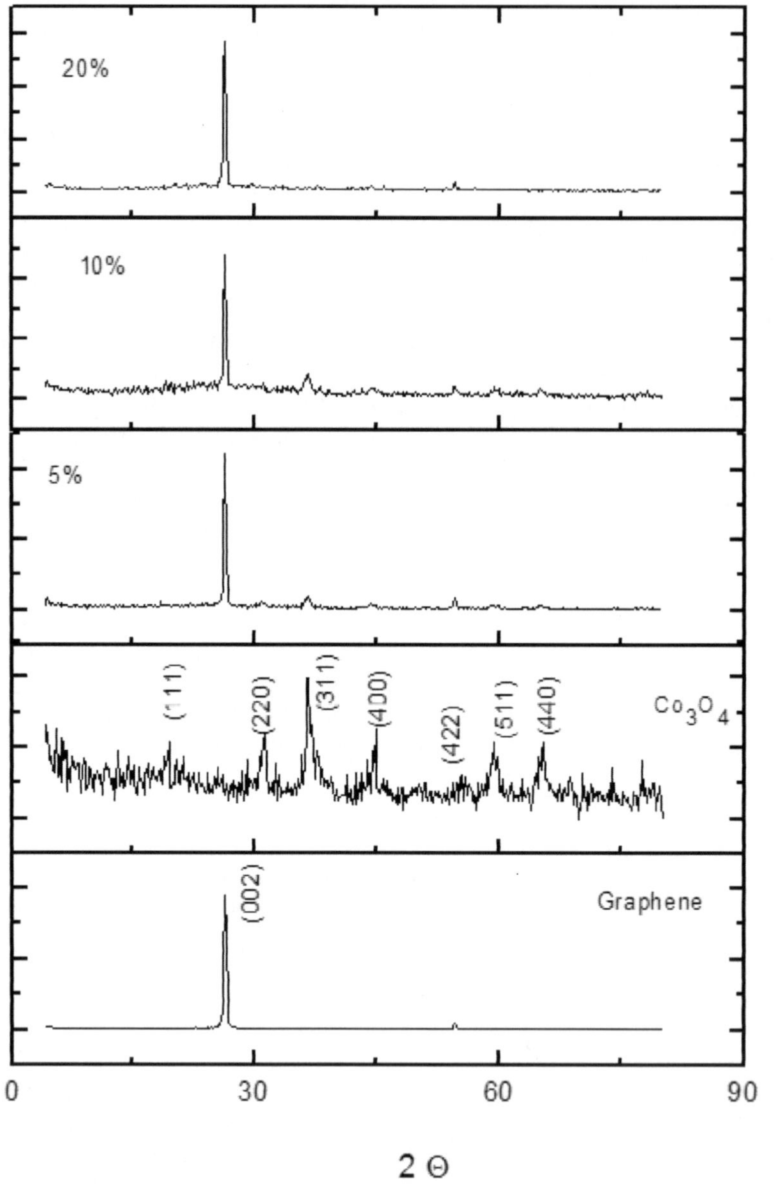

Figure1. XRD patterns of graphene, Co_3O_4 nanoparticles and Co_3O_4-graphene nanocomposite at different weight ratios of Co_3O_4 nanoparticles.

diffraction patterns of the cobalt oxide nanoparticles (Co_3O_4), graphene and composites with different molar ratios of Co_3O_4 nanoparticles display the reflections at 2θ = 19.018°, 31.177°, 36.884°, 44.817°, 55.729°, 59.314° and 65.472°. These peaks are associated with the crystallographic planes (111), (220), (311), (400), (422), (511) and (440) which agree with the typical spinel Co_3O_4 cubic structure [16]. The average crystallite size of cobalt oxide nanoparticles was estimated from Scherrer equation using the corrected full width at half maximum (FWHM). As reported in the literature [17], the Scherrer's equation is described as follows: $L = (k\lambda)/(\beta \cos\theta)$, Where L is the average crystallite size (nm), λ is the X-ray wavelength (0.1540593 nm), θ is the corresponding Bragg angle, $k = 0.9$ is the shape factor and β is the corrected FWHM.

Table 1. Structural parameters of graphene- Co_3O_4 nanocomposites

Sample	a (Å)	V (Å^3)	D_x (g/cm^3) of Co_3O_4	L_{avg}.(nm)
Pure Co_3O_4	8.0804	527.32	6.063	17.65
5% Co_3O_4	8.1248	536.34	5.960	19.06
10% Co_3O_4	8.1269	536.57	5.958	20.46
20% Co_3O_4	8.1301	537.36	5.949	24.06

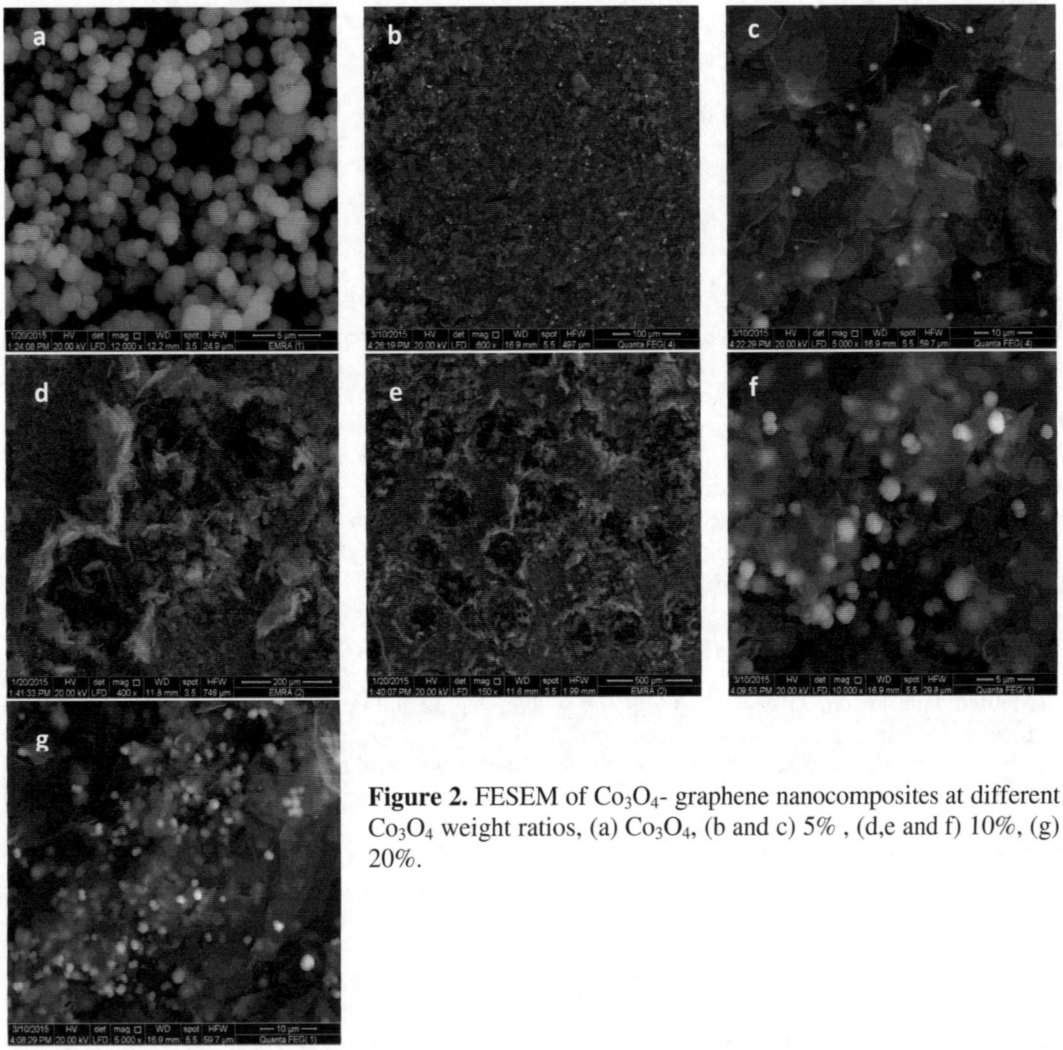

Figure 2. FESEM of Co_3O_4- graphene nanocomposites at different Co_3O_4 weight ratios, (a) Co_3O_4, (b and c) 5% , (d,e and f) 10%, (g) 20%.

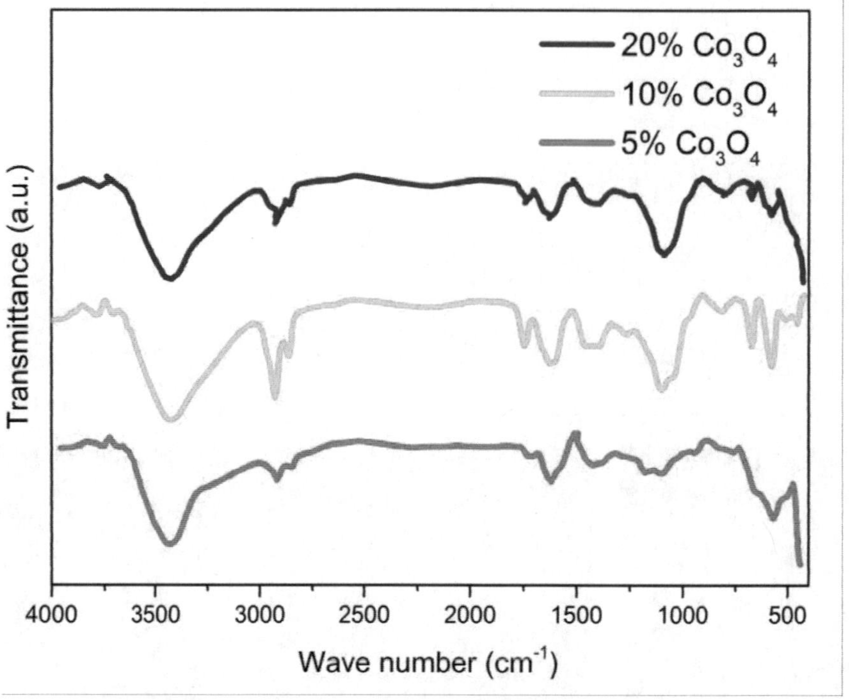

Figure 3. (a-c) FTIR spectra of Co_3O_4- graphene nanocomposites

The lattice parameters unit cell volume, theoretical density and average crystallite size for G-Co_3O_4 nanoparticles are reported in table 1.The data in the table assure that Co_3O_4 crystallite increased significantly in size when decorating graphene sheets, this means that the existence of the graphene layers didn't impede the crystal growth completely and Co_3O_4 nanoparticles were used successfully to prevent the stacking of graphene sheets and to offer an excellent conducting matrix and/ or template anchored with metal oxide nanoparticles. Moreover, the lattice parameter of the cubic unit cell as well as its volume increased with increasing Co_3O_4 weight percent in the nanocomposite. The lattice expansion; here in; depended strongly on the Co_3O_4 ratio which takes the upper hand in controlling the spacing between the graphene layers.

3.2. Field emission scanning electron microscopy (FESEM)

Figure 2 (a) illustrates the FESEM of Co_3O_4 nanoparticles. Nanospheres appeared homogenously distributed with uniform size, stacked together without significant coalescence despite the absence of surfactant. The observed grain size ranged from 90 to 100nm.

In Fig. (2:b,c) the sample with 5% Co_3O_4 nanoparticles morphology looks like homogenously distributed nanospheres anchored on the graphene nanosheets. The graphene nanosheets are well separated and the Co_3O_4 nanoparticles prevented their overlap.

From a closer look to the FESEM micrograph, it is obviously remarked that Co_3O_4 nanoparticles here didn't predominate due to their small content weight ratio. Graphene nanosheets appeared to be curved in a flower like morphology as clear in Figure 2 (d-f) representing 10% Co_3O_4. The layers of graphene appears as artichoke like carnations homogenously distributed. The nanosheets appeared to be as flower petals decorated with Co_3O_4 nanospheres. The micrographs give insight about more information useful in many applications. The graphene acted successfully as 2D template for enhanced crystal growth of Co_3O_4 nanospheres.

Herein and from the micrographs 20% of the Co_3O_4; Figure 2 (g) by increasing Co_3O_4 nanoparticles, they become more distributed and interesting features are clearly seen. The Co_3O_4 nanoparticles prevent the intercalation of the graphene and their overlay while the later helps in more homogenous dispersion of the former. This excellent morphology presented in a decorative manner where the graphene were arranged in flower like morphology and curved layers assure the strong interaction between Co_3O_4 and graphene nanosheets with good mechanical flexibility.

These nanocomposites are now recommended in energy storage applications owing to their large surface active area, since the anchoring of Co_3O_4 successed effectively to reduce the restacking of graphene sheets.

3.3 FTIR spectra

The infrared spectra of the produced Co_3O_4-graphene, nanocomposite samples, are presented in Figure 3 (a-c). The data show absorption bands at 668 and 571cm^{-1} which confirms the spinel structure of Co_3O_4. This absorption band is not clear in the sample (a) due to small amount of Co_3O_4 band (5%). The first peak is specifically assigned to the stretching vibration mode of $\nu_{Co\text{-}O}$ for Co^{2+}, while the band appeared at 571cm^{-1} is attributed to the Co-O bond for Co^{3+} [18,19]. The IR spectra, of the Co_3O_4 - graphene, nanocomposite samples exhibited a broad band at 3428, 3425 and 3424cm^{-1}, respectively corresponding to the frequency vibrations of (OH) group of the water molecules due to drying process where the temperature was 40° C only. The bands appeared at 1627, 1619 and 1626 cm^{-1} are assigned to the stretching vibration of $\nu_{C=O}$ groups for graphene- Co_3O_4, nanocomposite samples, respectively. The band intensities are certainly dependent on the Co_3O_4 content ratio in the nanocomposite.

3.4 Optical properties of Co_3O_4 - graphene nanocomposites

The optical absorption properties of the as-prepared Co_3O_4- graphene nanocomposite (20% weight ratio of Co_3O_4 was investigated at room temperature by UV–vis spectroscopy figure 4. When incorporating Co_3O_4 on to graphene surface, the small content 5% and 10% weight ratios didn't reveal absorbance owing to the excellent conductivity of graphene. At 20% Co_3O_4, the band gap was calculated to have two obvious absorption peaks.

The optical band gap E_g of the Co_3O_4- graphene nanocomposite (20%) was calculated using the equation (1) [20,21]:

$$\alpha = A(h\nu - E_g)^n / h\nu \qquad (1)$$

Where α is the linear absorption coefficient, $h\nu$ is the photon energy, A is a constant involving properties of the bands, and n is ½, 3/2, 2, and 3 for directly allowed, directly forbidden, indirectly allowed, and indirectly forbidden transitions, respectively.

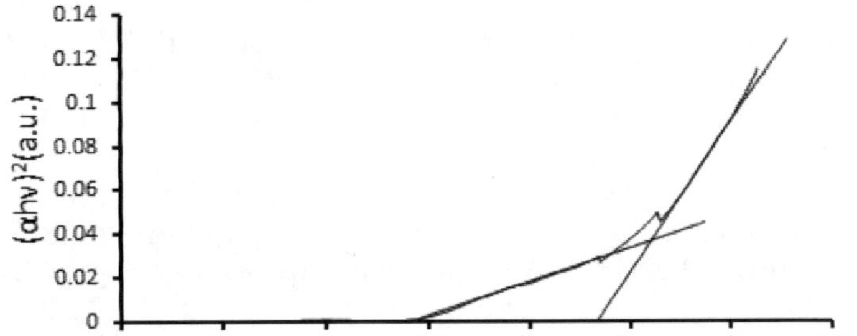

Figure 4. UV-Vis absorption spectra of 20% Co_3O_4- graphene nanocomposite.

The $(\alpha h\nu)^2$ versus hν curve is shown in the inset of figure 4. The value of hν extrapolated to $(Ah\nu)^2 = 0$ gives the absorption band gap energy. Two regions with a linear relationship are observed and giving two energy gap (E_g) values of 3.4 and 2.5 eV. It is known [22] that the band gap of Co_3O_4 nanoparticles reveal two E_g values of 3.2 and 2.3 eV . The band gap of 2.3 eV can be associated with the $O^{2-} \rightarrow Co^{2+}$ charge transfer process, while the 3.2 eV band gap relates to the $O^{2-} \rightarrow Co^{3+}$ charge transfer [23]. The increase in the band gap of the Co_3O_4-graphene nanocomposite (20%) may be ascribe to the quantum confinement effects or small size effects of the Co_3O_4 nanoparticles [24].

4. Conclusion

Co_3O_4 nanoparticles with average diameter of 90-100 nm were homogeneously adhered to the surface of graphene via a simplified sonochemical technique. The highly conductive matrix of graphene enabled the anchoring of Co_3O_4 nanoparticles in a well dispersed manner. The ratio of Co_3O_4 nanoparticles play the hero role in controlling the morphology of the graphene-Co_3O_4 composites, where we can tailor the physiochemical properties associated with the obtained morphology to achieve potential applications. Here, we concluded that graphene sheets acted as template for hosting metal oxide nanoparticles to be used in the optimization of the storage capacity of super capacitors. Moreover, increasing Co_3O_4 content could help in the use of these nanocomposites as modified anode for Li- ion batteries owing to the excellent conductivity offered by graphene itself. Our future work will be involved in exploring this methodology for the preparation of novel nanocomposites to offer multifunctional platform.

Acknowledgement

The authors would like to dedicate this work to the late Dr. M.A. Ahmed Professor of Nanotechnology and Materials Science, Cairo University, Egypt.

References

[1] Choi W, Lahiri I, Seelaboyina R and Kang Y S 2010 Synthesis of graphene and its applications: a review *Crit. Rev. Solid State Mater. Sci.* **35** 52–71
[2] Soldano C, Mahmood A and Dujardin E 2010 Production, properties and potential of graphene *Carbon N. Y.* **48** 2127–50
[3] Rao C N R, Sood A K, Voggu R and Subrahmanyam K S 2010 Some novel attributes of graphene *J. Phys. Chem. Lett.* **1** 572–80
[4] Kuilla T, Bhadra S, Yao D, Kim N H, Bose S and Lee J H 2010 Recent advances in graphene based polymer composites *Prog. Polym. Sci.* **35** 1350–75
[5] Fuhrer M S, Lau C N and MacDonald A H 2010 Graphene: materially better carbon *MRS Bull.* **35** 289–95
[6] Wu Z-S, Ren W, Wen L, Gao L, Zhao J, Chen Z, Zhou G, Li F and Cheng H-M 2010 Graphene Anchored with Co3O4 Nanoparticles as Anode of Lithium Ion Batteries with Enhanced Reversible Capacity and Cyclic Performance *ACS Nano* **4** 3187–94
[7] Chen S Q and Wang Y 2010 Microwave-assisted synthesis of a Co 3 O 4 – graphene sheet-on-sheet nanocomposite as a superior anode material for Li-ion batteries † *J. Mater. Chem.* **20** 9735–9
[8] Wang G, Liu J, Tang S, Li H and Cao D 2011 Cobalt oxide–graphene nanocomposite as anode materials for lithium-ion batteries *J. Solid State Electrochem.* **15** 2587–92
[9] Kim H, Seo D-H, Kim S-W, Kim J and Kang K 2011 Highly reversible Co3O4/graphene hybrid anode for lithium rechargeable batteries *Carbon N. Y.* **49** 326–32
[10] Li B, Cao H, Shao J, Li G, Qu M and Yin G 2011 Co3O4@graphene Composites as Anode Materials for High-Performance Lithium Ion Batteries *Inorg. Chem.* **50** 1628–32
[11] Zhu J, Sharma Y K, Zeng Z, Zhang X, Srinivasan M, Mhaisalkar S, Zhang H, Hng H H and Yan Q 2011 Cobalt Oxide Nanowall Arrays on Reduced Graphene Oxide Sheets with Controlled Phase, Grain Size, and Porosity for Li-Ion Battery Electrodes *J. Phys. Chem. C* **115** 8400–6
[12] Wang B, Wang Y, Park J, Ahn H and Wang G 2011 In situ synthesis of Co3O4/graphene nanocomposite material for lithium-ion batteries and supercapacitors with high capacity and supercapacitance *J. Alloys Compd.* **509** 7778–83
[13] Nassar M Y 2013 Size-controlled synthesis of CoCO3 and Co3O4 nanoparticles by free-

surfactant hydrothermal method *Mater. Lett.* **94** 112–5

[14] Zhou W, Liu J, Chen T, Tan K S, Jia X, Luo Z, Cong C, Yang H, Li C M and Yu T 2011 Fabrication of Co 3 O 4-reduced graphene oxide scrolls for high-performance supercapacitor electrodes *Phys. Chem. Chem. Phys.* **13** 14462–5

[15] Cobalt G À, Electrode O, Supercapacitor H and Detection E G 2012 3D Graphene À Cobalt Oxide Electrode for High-Performance Supercapacitor and Enzymeless Glucose Detection *ACS Nano* **6** 3206–13

[16] Wang X, Liu S, Wang H, Tu F, Fang D and Li Y 2012 Facile and green synthesis of Co3O4 nanoplates/graphene nanosheets composite for supercapacitor *J. Solid State Electrochem.* **16** 3593–602

[17] R. Jenkins R L S 1996 *Chemical Analysis: Introduction to X-ray Power Diffractometry* (John Willy and Sons, Inc., New York)

[18] Tatzber M, Stemmer M, Spiegel H, Katzlberger C, Haberhauer G and Gerzabek M H 2007 An alternative method to measure carbonate in soils by FT-IR spectroscopy *Environ. Chem. Lett.* **5** 9–12

[19] Šoptrajanov B, Stefov V, Kuzmanovski I, Jovanovski G, Lutz H D and Engelen B 2002 Very low H–O–H bending frequencies. IV. Fourier transform infrared spectra of synthetic dittmarite *J. Mol. Struct.* **613** 7–14

[20] Salavati-Niasari M, Davar F, Mazaheri M and Shaterian M 2008 Preparation of cobalt nanoparticles from [bis(salicylidene)cobalt(II)]–oleylamine complex by thermal decomposition *J. Magn. Magn. Mater.* **320** 575–8

[21] Herrero M, Benito P, Labajos F M and Rives V 2007 Nanosize cobalt oxide-containing catalysts obtained through microwave-assisted methods *Catal. Today* **128** 129–37

[22] Farhadi S, Safabakhsh J and Zaringhadam P 2013 Synthesis, characterization, and investigation of optical and magnetic properties of cobalt oxide (Co3O4) nanoparticles *J. Nanostructure Chem.* **3** 69

[23] He T, Chen D, Jiao X, Wang Y and Duan Y 2005 Solubility-controlled synthesis of high-quality Co3O4 nanocrystals *Chem. Mater.* **17** 4023–30

[24] Kumar R V, Diamant Y and Gedanken A 2000 Sonochemical synthesis and characterization of nanometer-size transition metal oxides from metal acetates *Chem. Mater.* **12** 2301–5

AUTHOR INDEX